KT-525-426

# This Book Will Blow Your Mind

## Journeys to the Extremes of Science

Edited by Frank Swain

**NewScientist**

First published in Great Britain in 2018 by John Murray (Publishers)
First published in the US in 2018 by Nicholas Brealey
An Hachette UK company

1

A CIP catalogue record for this title is available from the British Library

UK ISBN 978-1-473-62864-9
UK Ebook ISBN 978-1-473-62863-2
US ISBN 978-1-473-68503-1
US Ebook ISBN 978-1-473-68504-8

Typset in CelesteST by Palimpsest Book Production Limited,
Falkirk, Stirlingshire

Printed and bound by CPI Group (UK) Ltd, Croydon, CRO 4YY

John Murray policy is to use papers that are natural, renewable and
recyclable products and made from wood grown in sustainable forests.
The logging and manufacturing processes are expected to conform to
the environmental regulations of the country of origin.

John Murray (Publishers)
Carmelite House
50 Victoria Embankment
London EC4Y 0DZ

www.johnmurray.co.uk

Nicholas Brealey
Hachette Book Group
53 State Street
Boston MA 02109

www.nicholasbrealey.com

# Contents

## 3. Life at the Extremes

## 4. The Threadbare Fabric of Reality

# Introduction

Reality gets a bad rap. Too often it is cast as the sober canvas from which our imagination takes flight. Wild stories are called imaginative, while critics caution those who peddle them to 'be more realistic'. Worse still, we might say a tale 'lacks imagination', as if that were the only source of material that could excite us. The implication is clear: reality is humdrum.

Thankfully, nothing could be further from the truth.

We might be besieged by fake news and post-truth discourse, but even the most duplicitous of liars could never hope to invent a reality as marvellous, baroque or dazzling as our own. Who could dream up the millimetre-high mountain ranges that snake across the surface of neutron stars? Or the strange earthquake-warning lights that illuminate the sky above aching fault lines? Or suspect that buried deep below those fault lines, microbes that once shared the planet with dinosaurs are still plodding through their very long lives? The astonishing stories behind all these facts and much more are in your hands.

Few visit these dizzying corners of reality as frequently as the staff at *New Scientist*. Day in and out, we grapple with mind-boggling matters, always asking questions. Can time flow backwards? How do microbes survive outer space? Where do humans come from? And why do some physicists think the universe is a hologram?

This childlike questioning of reality does not mean we are averse to using our imaginations – *New Scientist*'s last book, *The Universe Next Door*, took a grand tour of the multiverse, stopping in at what-if worlds, alternative timelines and thought experiments. But this book is dedicated to reality in all its unimaginable complexity. Collected here are the stories we've published that left us bowled over or bamboozled. If you find yourself shaking your head in disbelief as you read them, know that we were doing the same when we worked on them. And yet, everything here is real – even if conclusive evidence for some bits still hovers tantalisingly out of reach.

We start with a peek at the strangeness of the everyday: why lightning bolts shouldn't exist, how the sun sometimes comes out at night, and the sea monsters that can appear from nowhere to break a ship in two. Then you're in the hot seat, as we find out why being human is the most incredible thing you'll ever experience (so long as we're not all holograms). Chapter three takes you on a wild safari of nature's strangest nooks, from the ecosystem above your head to the alien life beneath the seas.

And that is just the small stuff. Further chapters reveal a place on Earth where things fall up, hidden dimensions in space, the biggest number you can imagine and the small question of whether reality as we know it is, er, real.

This book will blow your mind and make you rethink what it even means to be you – but don't let that stop you reading it. As you will discover, the human mind is capable of incredible feats, so putting your blown mind back together at the end of it should be child's play. Really.

# 1 Seventeen Impossible Things Before Breakfast

You don't have to travel far to see things you didn't believe were possible. There are astonishing, unexplained events happening all around you. Nights as bright as day, hyper-computers that can think like us, exotic metals made with forbidden chemistry, and more. Join us as we start our expedition into the realm of the incredible. More than once you will think to yourself that these things simply cannot be true, but everything that follows is definitively documented – and totally mind-blowing.

# The riddle of the nocturnal sun

*What could be more sure than night following day? Yet before artificial lights blinded our sight, reports of nights as bright as day were common, discovers* **Rebecca Boyle**. *What lay behind the phenomenon was a mystery – until now.*

In the millennia before street lights and smartphones, humans could, on rare occasions, walk around on a moonless night and see clearly. Looking up, they could see broad luminous patches of light stretching across the sky, which brightened the heavens in all directions as though it were daylight. People could read without candlelight, view small details in their surroundings, and make out landscapes in the distance. It was as if the world were illuminated by a hidden night-time sun.

The existence of bright nights is well accepted, but their cause remains a mystery. Frustratingly, sightings have almost entirely faded away in the past few decades, making it seem that any hope of solving the riddle was dimming. Now, though, one man says he has seen the solution.

The earliest account of a bright night comes from Pliny the Elder, a Roman army commander who studied nature in his spare time. In his encyclopaedic *Natural History* of around AD 77, he wrote that the 'phenomenon commonly called a nocturnal sun . . . a light emanating from the sky at night' has been seen many times. In 1988, a French atmospheric scientist named Michel Hersé produced the definitive collection of accounts of bright nights, which documented similar stories from the past millennium and all over the world. In French, they were *nuit claires*, and in German *helle Nächte*.

But sightings have become rarer. The most recent may be from 22 and 23 August 2001 at the Leoncito Astronomical Complex high in the Argentinian Andes. During that event, Steven Smith of Boston University in Massachusetts and his colleagues reported a night sky that was 10 times brighter than normal.

There is an obvious reason why the frequency of reported bright nights might have fallen: it has to be dark for us to notice them, and these days, 99 per cent of people in Europe and North America sleep under an artificially lit sky.

Hersé's book suggests that about one bright night used to be observed every year, but aside from that reveals no obvious temporal or geographical pattern. 'I think you would have to have been in the right place at the right time, and in the right situation, to see one,' says John Barentine, an astronomer at the International Dark-Sky Association, which works to combat light pollution.

## Luminous smog

However, Barentine points to an interesting clue buried in nineteenth-century accounts. These frequently include a description of a 'luminous smog' in the air. Astronomers and maritime observers said the effect was distinct from auroras or the faint nocturnal glow known as the zodiacal light, a pyramid-shaped brightness produced when space dust reflects sunshine coming from below the horizon. This suggests there might be some sort of reflective haze hanging in the upper atmosphere.

Perhaps that could have been dust from volcanoes or meteors, says Barentine. Take the account of a diarist we know only as M. Toucher, writing near Paris on 30 June 1908. It is possibly no coincidence that this was the day of the Tunguska event, when a huge space rock exploded in the upper atmosphere over Siberia. People around the world reported a haze in the atmosphere for months afterwards, and light reflecting from the haze might explain why Toucher could write: 'At 22.30 . . . Very clear sky, full of stars which shine to the horizon. No moonlight. All the details of the landscape are visible.'

Despite Toucher's observation during a night when there was no moonlight, some have wondered whether bright nights could simply be cloudless nights with a full moon and bright stars. However, in 1909 L. Yntema, a doctoral student at the University of Groningen in the Netherlands, settled that question. After measuring the total amount of light from all the stars reaching Earth's surface, he found a discrepancy in the light on bright nights. That seemed to point to some sort of atmospheric phenomenon as their cause. Yntema called it 'Earthlight'.

## Rolling on the waves

So, are we sure that this isn't just a rare, mid-latitude aurora? That possibility was ruled out by Robert Strutt, son and heir to Lord Rayleigh, a physicist who, among other things, had discovered that the way gas molecules scatter light explains why the sky looks blue. The younger Rayleigh witnessed a bright night on 8 November 1929, and demonstrated that the light came from all directions. In an aurora, it typically comes in streaks.

Today, bright nights may have all but vanished, but we do have certain advantages over Rayleigh – satellites, for example. In the late 1980s, Gordon Shepherd of York University in Toronto, Canada, built a satellite instrument called WINDII, which could monitor waves of air as they rolled through the atmosphere. He soon found that these waves could pile up on top of one another to produce towers of pressurised air.

Along with the waves, Shepherd also studies how the chemical make-up of the atmosphere changes through the day. During daylight hours, ultraviolet radiation from the sun splits molecular oxygen into individual atoms. When the sun goes down, the atoms rejoin. This produces a small amount of light, called airglow.

Airglow is usually barely visible with the naked eye from Earth's surface, but looking at WINDII readings, which spanned from 1991 to 2004, Shepherd noticed the airglow emissions varied wildly from night to night, and from place to place.

In 2017, it occurred to him that air waves and airglow could be connected. The waves might force the oxygen into a higher concentration, he thought, creating a more intense glow that could explain bright nights. 'I don't know why it came to me, but I said, "Ah, that's the explanation",' he says.

To verify his suspicion, Shepherd first had to account for the sun's activity, which can affect the brightness of airglow too. He and his colleague Young-Min Cho went back through WINDII data for 1992 and 1996, when the sun's activity was at different levels. Cho wrote an algorithm that could search the data, discard any nights when there was an aurora and find times when waves might have piled up enough to produce a bright night. For both years, that analysis showed that the waves could have produced a bright night about 7 per cent of the time at any given spot on Earth.

That convinced them that the action of the waves was a greater influence on the airglow than increased solar activity. But it also indicated that you would get about 25 bright nights a year, which doesn't tally with Hersé's collected accounts. However, further analysis showed that stacked waves and a cloud-free night are not very likely to coincide at any given spot, reducing the expected frequency. 'I think that's pretty consistent with the historical record,' says Shepherd.

Shepherd, now in his eighties and retired, hasn't been able to link any of the suitably stacked waves with eyewitness reports of bright nights, partly because modern reports are so rare. To do so would be a neat confirmation, though, so he is looking into crowdsourcing a bright night. 'You could get together 500 to 1000 people via the web and cover different longitudes and potentially all nights of the year,' he says. Enough to brighten someone's night, anyway.

## Lightning shouldn't be possible

*You might have been told that lightning is created by highly charged particles in thunderclouds rubbing together, but the picture is far more mysterious, finds* **Stephen Battersby**.

One mystery is how thunderclouds become so highly charged. The best explanation is that collisions between small ice particles and heavier gobs of slush called graupel tend to transfer electrical charge, but the role of this process in real clouds is not proven.

An even bigger puzzle is how the huge current of a lightning bolt ever begins to flow when air is an electrical insulator. It is possible to make air break down to form a conducting plasma, but this requires a fearsomely intense electric field of more than a million volts per metre. Although meteorologists have sent hundreds of instrument-laden balloons and rockets into thunderclouds to test local conditions, the strongest fields they have seen are only about a tenth of that critical value.

Perhaps lightning needs some kind of catalyst to let fly? One theory is that cosmic rays are involved. These charged particles are probably generated by supernova explosions far away in the galaxy. A cosmic-ray proton can have enough energy to generate a cascade of relativistic particles when it collides with a molecule in the atmosphere. This cascade ionises the air, producing a conical shower of free electrons where a current might begin to flow.

Lightning often produces flashes of X-rays and gamma rays, and even beams of antimatter. These phenomena imply that some relativistic process is involved, but they don't prove that cosmic rays are the trigger.

To find out whether supernovae really are implicated, meteorologist William Beasley at the University of Oklahoma is collaborating with a team of physicists to develop a ground-based monitoring system. 'We are putting out cosmic-ray detectors along with our lightning mapping array to see if they coincide,' he says.

Even if they do, the puzzle will not be solved. The electrons liberated by a cosmic-ray shower are free for just a few microseconds – not long enough to maintain a large current. That is long enough, though, to boost the electric field in a cloud to a few hundred kilovolts per metre, says Beasley's colleague Danyal Petersen at the University of Nevada, Reno. A strong field might allow another process to kick in, as it can stretch raindrops inside a cloud into pointed, needle-like shapes. Like the point of a lightning conductor, these rain needles could enhance the local electric field, forming coronas of charged air. These could spread and merge, eventually forming an ionised path called a leader that can carry the full fury of a lightning bolt. Measuring the sequence of radio frequencies emitted by the whole process could test this theory.

## The metals made with forbidden chemistry

*From bronze to steel, alloys are the backbone of the modern world.* **James Mitchell Crow** *discovers how a recipe that shouldn't work is creating metal mixtures with totally unexpected abilities.*

One of the oldest shipwrecks ever discovered lies off the southern coast of Turkey. First glimpsed by a young sponge diver in 1982, it was carrying a curious cargo: 9 tonnes of copper and 1 tonne of tin. A curious cargo, that is, unless you know the recipe for bronze. The Uluburun wreck, named after a nearby town, dates from 1300 BC, smack in the middle of the Bronze Age.

Today, arguably, we live in the steel age, but the principle underlying our defining material remains the same. You take a pure metal and enhance it by adding a pinch of another element. One part tin to nine parts copper gives bronze; a smattering of carbon added to iron yields steel. This is the recipe for making alloys – materials whose strength, durability and workability make them the basis of everything in the modern world from cutlery to lamp posts to bridges.

But are traditional alloys the best we can do? Increasingly, metallurgists are questioning this received wisdom. By ripping up the millennia-old rulebook, they are making wild metallic mixtures where no single element dominates, and with it producing materials the likes of which we have never seen. With applications from nuclear fusion reactors to jet engines to basic chemistry and more, it's a rich new material seam to mine – and we've only just begun to scratch the surface.

There are good reasons why we have made alloys in the same way for so long. Because their layers of identical atoms can slip past each other easily, pure metals are soft. That's why gold panners can tell a pure nugget just by biting it. But introduce interloper atoms and you can disrupt that slipping, producing a tougher material. This idea has furnished us with many of the materials that underpin modern technology.

The heretical idea to throw out the established recipe came to researcher Jien-Wei Yeh in 1995. Driving across the Taiwan countryside to a meeting in Taipei, Yeh found himself pondering a problem ancient alloy-makers were already familiar with: there's invariably a point beyond which adding more of the alloying elements negates their benefits. At the atomic scale, the additive atoms start to form little clusters of metal-within-metal that make the material brittle.

Yeh's sudden thought was that the concept of entropy might provide a workaround. Entropy is a way of quantifying disorder in a system, and the rules of thermodynamics say that when something is more disordered it is more stable. So rather than make orderly alloys from one main element spiked with pinches of others, why not mix five, six or more elements? Swirl together enough elements in equal proportions, Yeh reasoned, and the resulting cocktail would be so disordered that there would be no chance for those crumble-inducing clusters to form.

Yeh was convinced he was on to something – so much so that he didn't go home after his meeting. Instead, he drove 80 kilometres back to his lab at National Tsing Hua University in Hsinchu City and immediately assigned a research student to work on the idea. He produced the first high-entropy alloy within one or two weeks, Yeh says. Within a year, they had made at least 40 different ones. From the very start, their properties looked promising: they were hard, tough and corrosion resistant.

But there was a problem. Yeh and colleagues couldn't be sure what they had made. Electron microscopy and firing beams of X-rays at a new alloy can usually tell us its structure, but Yeh's high-entropy creations were so unlike any other material that he had no reference data to make sense of what he was seeing. He spent the next eight years systematically making new materials, changing the composition bit by bit and comparing X-ray and microscopy scans until he had learned how to interpret them. In 2004, he was finally ready to unveil his idea.

## Treasure maps

It came at an opportune moment. 'In the past ten years, we've run out of ideas for new alloy bases for high-temperature structural metals,' says Dan Miracle, who works on jet turbines at the US Air Force Research Laboratory in Ohio. Once the initial scepticism had subsided, other researchers began to try their hand at making the new high-entropy alloys.

It was a daunting new research space. Imagine mixing equal amounts of five or more of the 60 or so elements commonly used in commercial materials in equal proportion. That gives you about 1040 possible combinations, says Dierk Raabe at the Max Planck Institute for Iron Research in

Dusseldorf. Relax the rules to allow 5 per cent variation in the proportions of the elements and you jump to 10,120 combinations. 'That's a vast untapped reservoir of possible compositions and properties we have not discovered yet,' says Raabe.

The trouble is, we could never even in principle make and test all these possible alloys, and it is not obvious where to begin looking for the ones with desirable properties. 'I think we're going to need treasure maps, theory-guided treasure maps, to show us where we can expect to find something really interesting,' Raabe says.

The internal structures of high-entropy alloys are so different, however, that we can't use existing theories to predict their behaviour. But even without a map, researchers have already made a few expeditions into uncharted territory – and come back with exotic new materials. 'We've seen some properties that are a little bit expected, and others that are totally unexpected,' says metallurgist Cem Tasan at the Massachusetts Institute of Technology.

One of those unexpected findings has to do with brittleness. All known alloys become more shatter-prone when cooled, but in 2014, Easo George at Ruhr University in Bochum, Germany, and his team found a high-entropy alloy of iron, manganese, nickel, cobalt and chromium that became less brittle the colder it got, right down to -200 °C.

Why the material bucks the trend is not yet clear, but it has already caught the eye of researchers hoping to make nuclear fusion a reality, says George. Taming fusion, the process that powers the sun, involves containing a cloud of charged particles at temperatures in excess of 100 million degrees. This in turn requires superconducting electromagnets that must be kept very cold – without breaking. 'You need high strength at cryogenic temperatures, and if it fails you need graceful rather than catastrophic failure,' says George. His alloy isn't magnetic, but it is made from elements with interesting magnetic properties, so similar alloys might combine magnetism with strength at low temperatures.

## Hard to crack

This is not the only cast-iron principle of materials science that looks a little more pliable with the advent of high-entropy alloys. There's also the idea that the harder something is, the more prone it is to crack when struck: think of a soft clay compared with a hard china teacup. In 2016, Raabe, Tasan and their colleagues stumbled on a high-entropy alloy containing iron, manganese, cobalt and chromium that ignores this rule. They think this behaviour must stem from multiple crack-preventing mechanisms at play, corresponding to various atomic rearrangements the material can adopt to absorb the force of an impact.

The chemical properties of some of these alloys are confounding researchers, too, making it all the more important to work out how to make the useful ones. One brute-force approach, adopted by Matthew Kramer and his team at the US Department of Energy's Ames Lab in Iowa, is simply to try out as many combinations as possible. They have developed a machine that can 3D print up to 30 rod-shaped alloy samples in an hour and test their physical properties automatically.

But what are we looking for? With more ingredients needed to make these new materials, some of them exotic and expensive, it's hard to see a high-entropy alloy replacing cheap, plentiful steel in buildings or bridges, for instance. But besides nuclear fusion reactors, there are plenty of niche environments where we would wish for better materials than we currently have.

High on the wish list are alloys that would turn waste heat – from a car exhaust, say, or laptop chip – into electrical energy. Materials that can perform this trick, known as thermoelectrics, spontaneously generate a current when one side of the material is hot and the other cold. That requires good electrical conductivity, but poor thermal conductivity to maintain the temperature difference. In regular materials electrical conductivity goes hand in hand with thermal conductivity, but early research hints high-entropy alloys might get around this logjam: their structural complexity could suppress heat flow through the material, while allowing electrons to whizz through unimpeded.

Miracle's employer, the US Air Force, also has some special material requirements. He is hunting for an alloy that can raise the operating temperature of a jet engine: the hotter you run a turbine, the more efficiently it performs. 'Getting better gas mileage out of our jet engines is way more important than lightening the load a little bit by making a lighter wing material,' he says.

A random search for this sort of material is unlikely to pay dividends – but we can narrow the field using what we already know. Miracle is concentrating on a cluster of elements in the centre of the periodic table known as the refractory metals. These elements, including molybdenum and niobium, have unusually high melting points of around 2500 °C, compared with 1455 °C for nickel, on which jet engine alloys are now based.

Pure niobium and molybdenum corrode too quickly to be useful, says Miracle. Mix them with other materials that offer corrosion resistance, however, and new possibilities might emerge. So far he has shown that high-entropy refractory metal alloys do indeed remain strong at temperatures beyond the wilting point of existing nickel superalloys. Corrosion resistance is one of the next properties to test.

And it's not necessarily the case that the alloy-making revolution will bypass all but the highest of high tech. Not, at least, if Kevin Laws has his way. At his lab at the University of New South Wales in Sydney, he has made high-entropy versions of those most ancient of alloys – bronze and brass. Both are so useful that they are still found in all sorts of everyday items, from keys and coins to bathroom taps. The main metal in conventional brasses and bronzes is copper, which is costly. Laws substitutes most of this with nickel, manganese, zinc and aluminium, a mixture that's less expensive overall and gives better performance. 'We've made them stronger, more corrosion resistant and cheaper,' says Laws. It looks set to be one of the first commercially produced high-entropy metals. Laws's team has already signed an agreement with Swiss alloy-makers Ampco Metal, which will probably use it first for making wear-resistant brass parts for industrial machines and potentially car engines.

Back when the Uluburun ship was sailing, alloys were used to make art, utensils and tools, as well as a new weapon that would help shape the jostling empires of the time: the sword. Who knows what technologies the new era of mixed-up metals will bring? Perhaps they will be just as world-changing.

## Four impossible things the laws of physics might allow

*Einstein's general theory of relativity is famous for its prediction of wormholes – shortcuts that might allow time travel by connecting different areas of space and time. Nobody's ever seen one, though, and debate rages over whether you could travel down one even if they did exist. While we wait for a visitor from the future to let us know,*

*Gilead Amit explores some other physical impossibilities that might already have been proved possible.*

## Perpetual motion machines

The idea of devices that can move and do other sorts of useful work with no external power has seduced some famous names over the centuries. Leonardo da Vinci worked on several designs involving spinning weights. Robert Boyle imagined a funnel that feeds itself. Blaise Pascal wisely abandoned the search and invented the roulette wheel instead.

Large-scale perpetual motion machines would break all sorts of physical laws, not least the cast-iron laws of thermo-dynamics. But Nobel Prize-winning theoretician Frank Wilczek's 'time crystals' – materials that eternally repeat in time with no external power source – seem to come close. Examples made in the lab in 2016 don't do any useful work, however, and so the quest continues.

## Teleporters

Ever wish that the ground would swallow you up and spit you back out somewhere far away? Strangely enough, there is nothing in the laws of physics to stop that happening. In his 2008 book *Physics of the Impossible*, physicist Michio Kaku calls teleportation a 'Class I Impossibility', meaning that the technology is theoretically feasible, and could even exist within our lifetimes.

In fact, teleporters already exist: not for whole human beings, but for subatomic particles. Quantum entanglement, the phenomenon that Albert Einstein called 'spooky action at a distance', allows information and quantum states to be trans-mitted apparently instantaneously across space. The first quantum teleportation experiments, carried out in 1997, involved one photon's quantum state being reconstructed in another photon tens of centimetres away. Today, the world quantum teleportation record stands at over 100 kilometres.

## Invisibility cloaks

Harry Potter's invisibility cloak is just one fictional example of magical garb that makes you disappear. But so-called metamaterials suggest a similar possibility in real life too.

The principle behind metamaterial cloaks is simple: waves of light bend around an object in your field of vision, much like water folds itself around a boulder in a stream. In practice, though, whole new nanostructured materials must be developed that can bend light in unfamiliar ways.

The first metamaterials were made in the lab in 2000, and basic cloaking devices soon followed. Cloaking has been ruled impossible for human-sized objects, but that's no great loss – even if it were possible, you would only be able to reroute specific wavelengths of light, making the cloaked object weirdly coloured and more conspicuous. Instead, similar cloaking principles might be used to divert seismic waves and shield entire cities from earthquakes.

## Matter married with antimatter

Normally, when matter comes into contact with its opposite, antimatter, both 'annihilate' in a sudden burst of energy. It's just lucky we live in a universe with a lot of matter and mysteriously little antimatter.

But then again, bizarrely, some matter might also be antimatter. So-called Majorana fermions would be their own antiparticles, capable of self-annihilating under the right conditions. Physicists have long suspected that neutrinos could fall in this category, although proving that means spotting some of the rarest processes in the universe in action, that happen perhaps once in 100 trillion trillion years.

Meanwhile, there are persistent reports we've made something similar in the lab. When an electron is torn out of a superconductor, a hole is left behind that acts like a positively charged particle with exactly the same mass. If the two are manipulated in just the right way, they can be made to act like Majorana particles.

## The clinic offering young blood to cure ageing

*In 2016, a California start-up began offering $8000 blood transfusions to people who hope they can turn back the clock. Is it safe, and will it work?* **Sally Adee** *rolls up her sleeve for a dose of the elixir of youth.*

I discover the French bistro tucked into a strip mall that wraps around a parking lot it shares with a hardware shop and a barber. The man I am meeting has asked to come here instead of the clinic because he doesn't want to do an interview while being transfused.

I am led to a booth where I find him drinking a glass of wine, wearing the blazer-and-T-shirt uniform common among venture capitalists. His youthful looks have an enhanced, slightly uncanny cast, but I am still shocked when he tells me he is 65. To protect his privacy, he chooses to be identified for this article as JR.

JR is a minor celebrity in these parts. It is the fifth time this year that he has flown in from Atlanta to have the treatment. Monterey doesn't get a lot of traffic from people like him. You might think of the Californian coast as a homogeneous stretch of sun-drenched sand and rich people. But things change around the middle of the state. Driving there from San Francisco, you quickly lose the sun behind a permanent blanket of fog. The central coast is long and flat, with glimpses of iron grey choppy waves behind squat buildings.

So it's a bit odd that this is the epicentre of a phenomenon rocking Silicon Valley: young blood treatments. JR is one of about 100 people who have each paid $8000 to join a controversial trial, offering them infusions of blood plasma from donors aged between 16 and 25 in a bid to turn back the clock. Participants have come from much further afield, including Russia and Australia.

It's not hard to see why. After a spate of trials showed astonishing rejuvenation in old mice, the notion of filling your veins with the blood of the young has gone from creaky vampire myth to the latest tool in Silicon Valley's quest to 'disrupt death'.

Now start-ups, universities and pharmaceutical companies are clambering to commercialise the potential of young blood. Venture capitalists and high-level hospital executives are rumoured to be partaking behind the scenes. The idea's popularity is sparking fears of red markets and a dystopian future in which the old steal youth from the young, and no longer just metaphorically.

Scratch beneath the hype, however, and we may have been looking at young blood the wrong way round. Within a few years, new insights could usher in a safer, more effective way for blood to stop the inevitable declines of ageing.

## Mystery ingredients

Vampire tales aside, we have suspected since the mid-nineteenth century that young blood has rejuvenating powers, thanks to a grim surgical technique known as parabiosis. Scientists would stitch together two animals, usually rats – like twins conjoined only at the skin – and wait a week for capillaries to form and fuse their blood supplies. The new plumbing seemed to change the old rats, making them physically and cognitively resemble their younger partners. By 1972, research began to suggest that after being conjoined, old rats even lived longer.

In the early 2000s, researchers at Stanford University in California revisited the technique, this time with a view to reversing specific ailments of ageing. They damaged the livers and muscles of old mice before connecting each one up to an undamaged mouse. Those with young partners healed well. Those with old partners did not. Similar results emerged with regard to heart health, then age-related cognitive declines.

What was it in blood that was having these rejuvenating effects? The prime suspect seems to be plasma, the yellow liquid that gets separated out after donation. Components like red blood cells are used for medical transfusions, but the plasma often goes spare.

Plasma is rich in all sorts of proteins and other compounds, which could hold the key to what makes young people young and old people old. Not that we know what all these components are. But we do know that their amounts and ratios change as we age. For example, old blood has higher levels of inflammatory compounds that damage tissues they reach. Inflammation has been linked to cancer, heart disease and depression. Younger blood, by contrast, is characterised by a higher concentration of stimulating and restorative factors.

An amazing discovery, but to be medically relevant, we must deliver young blood without having to stitch pensioners to 20 year olds. So in 2014, a team led by Tony Wyss-Coray, a neuroscientist at Stanford University, injected middle-aged mice with plasma from young mice. Sure enough, after three weeks they had anatomical improvements in the brain and a cognitive boost, compared with mice given a placebo. Every other system they tested fared similarly.

The plasma didn't even need to come from the same species – old mice became just as sprightly when the injection came from young humans. 'We saw these astounding effects,' Wyss-Coray told *New Scientist* in 2014. 'The human blood had beneficial effects on every organ we've studied so far.'

Wyss-Coray had the proof he needed to start a human trial. In October 2014, his start-up, Alkahest, began recruiting participants for a trial at Stanford School of Medicine, using young blood in people with late-stage Alzheimer's disease. The following year, Bundang CHA General Hospital in South Korea launched a gold-standard trial to compare the anti-ageing effects of cord blood, young plasma and placebo on markers of frailty in ageing. Both trials were met with enthusiasm. Wyss-Coray was invited to give lectures, including at the World Economic Forum and a TED talk.

Then there's the Ambrosia trial, which JR is taking part in. Ambrosia is a start-up headquartered in Washington DC. The trial didn't need regulatory approval because plasma is already a standard treatment to replace missing proteins in people with rare genetic diseases. And there's no placebo arm to it. All you need to join is a date of birth that makes you over 35 – and a spare $8000.

For your money, you are infused with 2 litres of plasma left over from young people who have donated to blood centres. Unlike the trials looking at young blood's effects on specific diseases, Ambrosia has a softer target: the general malaise of being old. In addition to measuring changes in about 100 biomarkers in blood, the firm is also 'looking for general improvements', says Jesse Karmazin, who runs the start-up.

The methodology falls short of the normal standards of scientific rigour, so it's unsurprising that scientists and ethicists have accused Karmazin's team of taking advantage of public excitement around the idea. 'I don't think the Ambrosia trial can be called a trial at all, since they treat healthy people and they have no clear read-outs,' Wyss-Coray says.

This makes any findings virtually unpublishable, which may explain why Karmazin announced his first results to a room full of technologists at the Silicon Valley Code Conference in early 2017 instead of at a medical conference or in a journal. The numbers were as unverifiable as they were impressive: one month after treatment, 70 participants saw reductions in blood factors associated with risk of cancer, Alzheimer's disease and heart disease, and reductions in cholesterol were on par with those from statin therapy.

Karmazin says this could explain his observations during the trial: a woman with chronic fatigue syndrome is now able to get out of bed and live normally; another participant, who had early stage Alzheimer's when he enrolled, no longer meets the clinical criteria for having the disease.

'Whatever is in young blood is causing changes that appear to make the ageing process reverse,' Karmazin told me. Even healthy participants 'just have more energy'.

JR agrees. 'I do feel it a bit,' he says. 'I am starting to run again now.' However, although Karmazin says that the effects in the blood are identical regardless of participant age, JR says his 39-year-old girlfriend feels no different after two treatments. As for his youthful looks, JR says he tries out many of the therapies his company invests in.

Not that any of this should be taken at face value. Many of Ambrosia's claimed improvements could be down to the placebo effect. Even so, the numbers are proof enough for Karmazin. He originally aimed to recruit 600 participants.

However, the results have made him so optimistic that he is expanding the business. When I travelled to Monterey in June 2017, Karmazin had just opened his third clinic, and thanks to recent infusions of investor cash, he is planning a total of six in the US in 2018.

I ended up in Monterey because, for the past year, I've been preparing myself to enrol in the trial too.

As it did with much of the public, the idea of the glittering Silicon Valley 'blood spa' captured my imagination. The reality of this clinic is somewhat different, though. The one-storey building shares an intersection with a flaking self-storage facility and a pockmarked parking lot.

The interior is also modest; patients pass through a wood-panelled kitchen on their way to the reception. In the main room, a row of armchairs, each with an IV drip stand, faces a window overlooking washed-out scenery that culminates with the Pacific Ocean, barely discernible under the fog. Most of the elderly clients occupying those chairs are not getting plasma, but IV fluids.

When I visit, the trial is being run by Craig Wright, Karmazin's erstwhile partner. Karmazin has a medical degree but not a licence to practice, so he teamed up with Wright, an immunologist formerly at Walter Reed Army Medical Center in Washington DC, who is licensed to run one of the West Coast's few non-hospital affiliated infusion clinics.

Although at 67 he could be retired, Wright maintains the clinic to look after his patients. 'Healthcare in this country doesn't give a crap about old people,' he says. One of his clients has dementia that used to regularly put him into the emergency room for dehydration. Another, after several rounds battling lymphoma, had a compromised immune system that left her struggling with recurring infections.

Eventually Wright enrolled her in the Ambrosia trial. Her

infections went away. But by the time I get to the clinic, I am rethinking whether I want to go through with this. Wright and I go into his office so he can give me the consent form. I tell him I'm starting to reconsider, and I am surprised to find he doesn't try to talk me round. 'You need to think long and hard before you do this,' he says. For some of his older, sicker patients, plasma has proved beneficial. For younger would-be participants like me, however, he lists a litany of potential side effects.

Risks commonly associated with plasma transfusion include transfusion-related acute lung injury, which is fatal; transfusion-associated circulatory overload; and allergic reactions. Rare complications include catching an infectious disease: blood products carry a greater than 1 in a million chance of HIV transmission. That's too risky for JR, who tells me that before every treatment he takes a dose of the HIV prophylactic PrEP.

Karmazin had previously assured me that none of the risks associated with plasma transfusion exceed 1 or 2 per cent, a statistic borne out in the trial: in his Code presentation, he told the room that 'none' of his participants had reported any negative effects. 'Not one.'

## Complications

But when I meet up with JR and Wright on the second day of my visit, both are visibly shaken. A participant had arrived earlier that day from Moscow to get the infusion. As he started on his second unit of plasma, the man had an anaphylactic reaction. His face and tongue swelled up, and he developed a rash all over his body. 'Even the whites of his eyes turned red. He was in a lot of trouble,' says JR. Wright administered emergency treatment to stabilise him and sent him back to his hotel.

I am astonished that my visit has coincided with the first complication of the treatment. There's an uncomfortable silence as JR and Wright exchange glances. 'It's not the first one,' says Wright. When I press him for more information, he demurs. 'You'll have to talk to Jesse.'

When I call Karmazin, he clarifies that there was also an eyelid rash and a case of pneumonia that was probably already in place before the patient got the treatment. But in later conversations with Wright, he tells me of worse cases. Without published data, it's impossible to get to the bottom of it.

Either way, those are the known knowns. There are also known unknowns associated with injecting material from someone who is genetically different to yourself, says Irina Conboy, a co-author on the early Stanford work that put young blood on the map, and now at the University of California, Berkeley.

There could be risks of developing autoimmune disorders. And some fear that pumping stimulating proteins into people for years could lead to cancer. 'If you keep infusing blood, the risk of reactions goes up,' says Dobri Kiprov, an immunologist at California Pacific Medical Center in San Francisco. 'Many of these people are just eager to get younger – they don't have a particular disease, so it's not justified.'

It's easy to criticise Ambrosia for charging people to receive an unproven procedure, but there is also no evidence yet that the other trials are yielding anything more promising. When Alkahest presented results from its Alzheimer's trial at a conference in November 2017, it revealed that the control group had been abandoned, something critics said made any positive findings irrelevant.

## The man who wants to transplant human heads

*Neurosurgeon Sergio Canavero shocked the world when he announced plans to perform the first ever human head transplant. The maverick surgeon likens himself to Peter Parker and Victor Frankenstein, and in 2017 told* **Nic Fleming** *that a successful procedure was imminent. What happened next?*

There's a story doing the rounds about Sergio Canavero. One day, as a 9 year old, he sought refuge from playground bullies in the pages of a superhero comic. 'I myself have surgically rejoined severed neurolinkages,' declares Dr Strange in the November 1976 issue of *Marvel Team-Up*. The brilliant but egotistical fictional surgeon continues: 'The nerve endings have been fused – the healing process begun.' The young Canavero was captivated. Four decades on, he announced plans for the world's first human head transplant . . .

If this story sounds a little too neat to be true, that's because it probably is. The Italian neurosurgeon, who describes himself as a 'big comics nut', says he didn't read that issue until adulthood. He claims the publication that reported the comic as inspiration for his work was mistaken.

Whatever the true picture, Canavero does not flinch at comparisons with fictional characters. Quite the opposite, he encourages them. He sees a lot of himself in Peter Parker – aka Spider-Man – who, as a nerdy student, was bullied by classmates and shunned by girls. After dismissing the *Marvel Team-Up* story, he sends me PDFs of the relevant comic frames, along with a screen grab of a doctor discussing fusing severed spinal cords from the 2016 film *Dr Strange*. Canavero says he was the inspiration for that scene. 'I have good ties with Hollywood and I can tell you for a fact that line came out of my book.'

Since 2013, Canavero has been promoting the idea that head transplants – better understood as body transplants – are feasible, and should be offered to people with conditions involving muscular and nerve degeneration, for example. The response, in the West in particular, has ranged from disbelief and opposition to the questioning of his motives and scientific credibility.

Unperturbed, in 2015 Canavero agreed to help set up a team to carry out the procedure in China, working with Xiao-Ping Ren, an orthopaedic surgeon at Harbin Medical University who helped with one of the first hand transplants in 1999. In mid-2017, Canavero claimed several papers supporting the feasibility of human head transplantation will soon be published, that an operation could go ahead by the end of the year. (Spoiler: it did, although perhaps disappointingly, it was performed on a pair of corpses.)

'The team in China is ready to roll,' says Canavero, who worked as a neurosurgeon at Turin University Hospital in Italy until 2015. 'All the preclinical and clinical studies have been conducted successfully.' Much of this work will not be published, he says, but insists that what will be published 'will be more than enough to show where China stands'. The precise date of a transplant attempt depends on finding a donor of the right height, build and complexion, he says. 'The problem now is only organisational.'

Canavero calls his proposed procedure the head anastomosis venture, or HEAVEN. He says it would begin by cooling the donor body and recipient's head to delay tissue death. The heads would be detached and the donor body attached to the recipient's head. Polyethylene glycol (PEG) would help fuse the cords by encouraging the fat in adjoining cells to mesh together. Stimulation from implanted electrodes would help to strengthen nerve connections. That's the plan, anyway.

Canavero doesn't agree with mainstream medical thinking that movement below the neck depends primarily on bundles of long-range nerve fibres in the spinal cord. Inspired by research dating back to the first half of the twentieth century, he believes a person undergoing a head transplant could regain close to full movement after those nerves have been severed, thanks to the regeneration and fusing of short-range nerve fibres that are part of an additional, interconnected network of cells in the spinal cord called propriospinal neurons. Think of a fire brigade passing buckets of water along a line if their main hose has been severed.

C-Yoon Kim of Konkuk University in Seoul, South Korea, is part of Canavero's group and has led animal experiments that use PEG to encourage regrowth of severed spinal cords. In a paper published in 2016, Kim claimed that five of eight mice whose spinal cords were severed and treated with PEG regained some movement after four weeks, while none of the mice in the control group did. In a paper published in 2016, Ren's team reported a similar experiment, claiming that five of nine PEG-treated mice 'regained independent ambu-lation, with two basically normal'. And in a widely criticised experiment in 2016, Kim's team claimed that a beagle with around 90 per cent of its spinal cord cut at the neck regained 90 per cent of its motor function within three weeks, with the aid of PEG. Critics pointed out there was no control in the study and no data proving the degree of damage to the dog's spinal cord.

Ren and Canavero also published a paper in the journal *Surgery*, in which they claimed to have carried out a monkey head transplant, described as successful on the basis of 'unpublished observations'. 'There's not such a big difference

between animals and humans – the basic functions, physiology and the possibility of recovery are the same,' says Kim.
'I believe we will succeed in the human operation in the near
future.'

Many scientists refuse to comment on the record about
Canavero's work, often because they have serious doubts
about the published papers. Most who have commented say
a successful human head transplant is far from feasible. 'There
is no way that I know of, that has been published, that would
allow fusion of the spinal cord,' says José Oberholzer, director
of the Charles O. Stickler Transplant Center at the University
of Virginia, Charlottesville. 'The rest, including reconnecting
the blood vessels and the airways, is theoretically possible
but very challenging, and carries major risks of complications
from leaks and immune system rejection.'

Others have used words such as 'charlatan' and 'self-
promoter'. Canavero dismisses his critics as too short-sighted
to share his vision. 'The history of science is full of people
being called nuts and then going on to prove their point,' he
says. 'The academe tried to destroy me, they tried everything
to stop me, to slander me, but they failed. It's no problem: I
practice ju-jitsu and the mindset is let your enemy come to
you, and then exploit his momentum to bring him to the
ground.'

Canavero claims the media has failed to report his work
properly. And in his efforts to seek favourable coverage he
tries tactics more often employed by spin doctors. For
example, he offers me 'exclusive' early access to as-yet unpublished papers if I interview specified people, avoid others
who have criticised his work, show him my article in advance
and we agree 'mutually acceptable' terms. I politely refuse
his offer.

Canavero, now 52, does not want to talk about his early

years, but the few details he does give me could help explain why he self-identifies as an outsider. He describes growing up as an only child in a poor neighbourhood of Turin in a 'difficult family situation'. He says he was bullied by classmates because he was bright, which made him resentful. 'I'm a loner, I'm a maverick,' he says. 'I believe the initial suffering when I was a child had a lot to do with it.'

When I ask Canavero about what motivates him, he does not mention helping patients. 'The goal is to understand the nature of consciousness and to answer the basic question of what happens when we die. I believe consciousness is not generated in the brain, which merely acts as a filter.'

Canavero believes the mind and body are separate and that so-called near-death experiences support this view. He wants to show that those who dismiss these reports as hallucinations are wrong.

'My idea is to generate near-death experiences,' he says. 'When you detach a head or a brain, the brain is cooled so there is no electrical activity and no blood flow, that brain is clinically dead. Patients will tell us about their near-death experiences and we will know for a fact that they could not have been generated by the dying brain.' That will change how we consider ourselves as human creatures, he says. 'That will trigger the greatest revolution ever. That is the final goal, the real goal.'

'Of course, there is another goal,' he continues, 'which is life extension.' The book Canavero says inspired Hollywood's scriptwriters is his 2014 work *Head Transplantation and the Quest for Immortality*, in which he predicted the transfer of human brains to artificial bodies 'possibly by 2025'. He also claims to be assembling a team to perform a human brain transplant, and that this procedure will happen within three years.

Many believe the chances are remote that Canavero's aspirations will become reality any time soon. 'The complexity of what he is proposing is enormous,' says neurophysiologist Peter Ellaway, an emeritus professor at Imperial College London. 'I think his aims are almost in the realm of science fiction. In my opinion this will never happen.'

Others fear indirect downsides to Canavero's actions, even if no head transplant is ever attempted. 'Being a transplant donor is one of the very few purely altruistic things a human can do,' says Oberholzer. 'When someone wants to be the hero by putting on this audacious show, people may be put off donating their organs or bodies if they think doctors are going to do something crazy like this.'

Superheroes aren't the only fictional characters Canavero likens himself to. Victor Frankenstein is another. When I confess I haven't read Mary Shelley's *Frankenstein*, he emails me quotes from the novel. Such as: 'Whence, I often asked myself, did the principle of life proceed? [. . .] I was surprised, that among so many men of genius who had directed their inquiries towards the same science, that I alone should be reserved to discover so astonishing a secret.'

Dr Strange-meets-Victor Frankenstein, with a dash of Spider-Man, might make for an entertaining science fiction character. And Canavero certainly loves to entertain, playing on his renegade scientist image. 'Are you sitting tight?' he asks, in full showman mode, at the start of a recording of his 2014 TEDxLimassol talk. 'I'm about to give you one hell of a ride.' He smiles at his audience's nervous laughter. But fast-forward to today, with the scene shifting from the stage to the operating theatre, and the Canavero carnival risks turning into a horror show. With real-life patients at stake, no one's laughing any more.

## Weird earthquake warning lights

*For millennia, people have reported strange, baleful lights appearing before and during earthquakes.* **Stephen Battersby** *investigates the origin of these ill portents.*

Glowing orbs that drift through the air; blue-white sheets of light; sparks and flashes and flames licking up from the ground . . . all may be signs of disaster to come. In 1746, the flames dancing on San Lorenzo Island in Peru impressed prison governor Manuel Romero so much that he briefly released the detainees to let them watch. Three weeks later a huge quake hit nearby Lima and a tsunami washed 5 kilometres inland.

There is plenty of photographic evidence for earthquake lights. They tend to accompany large quakes – with magnitudes above 6 – centred at fairly shallow points in Earth's crust. It is not clear how the lights are produced, but Friedmann Freund of NASA's Ames Research Center in Moffett Field, California, thinks that when rocks in the crust are squeezed, chemical bonds break to produce a pulse of electrical charge that travels up to the surface. 'The rocks become like a battery and produce an enormous amount of electric power,' he says.

This process only generates a low voltage, but Freund thinks that the charge forms an ultra-thin layer at the surface. Since the charge is concentrated over a small distance, it would create a strong electric field, perhaps enough to ionise the air and create a luminous discharge that travels up, away from the ground – explaining the orbs, flames and aurora-like sheets of light.

Freund does not know why the charge should form such a thin layer, or how the wave of ionisation is maintained for any distance through the air. But experiments are encouraging: crushing rocks in the lab produces electric charge and flashes of light. And low-frequency radio waves have been measured in earthquake zones, suggesting that there are currents underground.

As earthquake lights are so rare, it is hard to show that ionisation and the emission of radio waves coincide with them. Freund has yet to secure funding for a network of cameras and a data-processing centre to monitor such events, but he hopes that such a system, along with satellite imagery, could one day provide earthquake warnings akin to weather forecasts.

Lights may not be the only aerial omens of impending doom. In 2004, a curious linear gap appeared in the clouds above a fault line in Iran. An earthquake followed 69 days later. The gap opened again in 2005, and this time an earthquake followed after six days. Two Chinese geophysicists, Guangmeng Guo and Bin Wang, have suggested that hot gas escaping from the fault might cut through the clouds.

## The computer that goes beyond logic

*For 75 years, computers have worked within limits defined by Alan Turing.* **Michael Brooks** *reveals how work has now begun to fulfil his prophecy of a machine that can solve the unsolvable.*

He called it the 'oracle'. But in his PhD thesis of 1938, Alan Turing specified no further what shape it might take. Perhaps that is fair enough: aged just 26, the British mathematician had already lit the fuse of a revolution. His blueprint for a universal computing machine, published two years earlier, set the specs for every computer that followed, from the humblest pocket calculator to the mightiest supercomputer – via laptop, smartphone and all points in between.

So absorbed have we been in exploring this rich and varied legacy, and transforming our world with the machines and applications that built on it, that we have rather overlooked the oracle. Turing had shown with his universal machine that any regular computer would have inescapable limitations. With the oracle, he showed how you might smash through them.

In his short life, Turing never tried to turn the oracle into reality. Perhaps with good reason: most computer scientists believe anything approximating an oracle machine would soon fall foul of fundamental restrictions on how information and energy flow in the universe. You could never actually make one.

In a laboratory in Springfield, Missouri, two researchers are now seeking to prove the sceptics wrong. Building on theoretical and experimental advances of the past two decades, Emmett Redd and Steven Younger of Missouri State University think a 'super-Turing' computer is within our grasp. With it, they hope, could come insights not just into the limits of computation in the cosmos, but into the most intriguingly powerful computer we know of within it: the human brain.

Computers as we know them are in essence very capable, rigorous and efficient renderings of what we humans might be capable of if given precise instructions, a high boredom threshold and a limitless supply of paper and pencils. They excel at successive additions, multiplications, logical decisions, if $x$ then $y$, that sort of thing. Indeed, the first 'computers' were young researchers employed by astronomers for the tedious and time-consuming task of working out the orbits of comets or calculating the brightness cycles of variable stars.

A universal computing machine – often known simply as a Turing machine – does the same things, only without the tedium. 'Electronic computers are intended to carry out any definite rule-of-thumb process which could have been done by a human operator working in a disciplined but unintelligent manner,' as Turing himself wrote in the programmer's handbook for the University of Manchester's Mark II computer in 1950.

So computers have their blind spots just as we do. No matter how disciplined, well-schooled or patient we are, certain questions defy our logic. What is the truth of the statement, 'This statement is false'? You can spend a lifetime grumbling over the answer, as many a philosopher has. In 1931, the mathematician Kurt Gödel demonstrated that this problem was universal with his infamous incompleteness theorems, showing that any system of logical axioms would always contain such unprovable statements.

Similarly, as Turing showed, a universal computer built on logic alone always encounters 'undecidable' problems that never yield straight answers, no matter how much processor power you throw at them. One example is the halting problem. A computer can never tell if any program will run to the end, or get stuck in some infinite loop or at some instruction, without trying out the program first – and possibly getting stuck. The 'blue screen of death' feared by many a PC user is just one consequence of this fundamental undecidability.

An oracle, as Turing envisaged it, was essentially a black box whose unspecified contents would be able to solve undecidable problems. An 'O-machine', he proposed, would exploit whatever was in this black box to go beyond the bounds of conventional human logic – and so surpass the abilities of every computer ever built.

That is as far as he went in 1938. 'Turing realised that models for more powerful computing machines may exist,' says Younger. 'But he did not present any super-Turing computational models.'

Over two decades ago, Hava Siegelmann came up with one by accident. In the early 1990s, she was working on her PhD in computer science at Rutgers University in Piscataway, New Jersey, just a 40-minute drive from Princeton, where Turing had presented his thesis. Her subject was neural networks, circuits designed to mimic the human brain and the myriad neurons connected by synapses that realise its unparalleled computing power. In a neural net, many simple processors are wired together so that the output of one can act as the input of others. These inputs are weighted to have more or less influence, and the idea is that the network 'talks' to itself, using its outputs to alter its input weightings until it is performing its task optimally – in effect, learning as it goes along, just as the brain does. Neural nets have scored some notable successes performing tasks that cannot easily be reduced to a set of straightforward instructions, from reading medical scans and diagnosing illnesses to driving cars.

Siegelmann's initial aim was to prove theoretically the limits of neural networks: to show that, for all their flexibility, they could never have the full logical capabilities of a conventional Turing machine. She failed time and again. Eventually, she proved the reverse. One of the hallmarks of a Turing machine is that it is incapable of generating true randomness. By weighting a network with the infinite, non-repeating number strings of irrational numbers such as pi, Siegelmann showed you could, in theory, make it super-Turing. In 1993, she even showed how such a network could solve the halting problem.

Her fellow computer scientists met the idea with coolness, and in some cases downright hostility. Various ideas had been floated for 'hypercomputers' that might exploit exotic physics to go super-Turing, but they always seemed to lie on a scale from implausible to utterly wacky. Siegelmann eventually published her proof in 1995, but she soon lost interest, too. 'I believed it was mathematics only, and I wanted to do something practical,' she says. 'I turned down giving any more talks on super-Turing computation. I told everyone, "I'm out of this field now".'

Redd and Younger had been aware of Siegelmann's work for a decade before they realised that their own research was heading in the same direction. In 2010, they were building neural networks using analogue inputs that, unlike the conventional digital code of 0 (current on) and 1 (current off), can take a whole range of values between fully off and fully on. There was more than a whiff of Siegelmann's endless irrational numbers in there. 'There is an infinite number of numbers between 0 and 1,' says Redd.

## Powered by chaos

In 2011, they approached Siegelmann, by then director of the Biologically Inspired Neural & Dynamical Systems lab at the University of Massachusetts in Amherst, to see if she might be interested in a collaboration. She said yes. As it happened, she had recently started thinking about the problem again, and was beginning to see how irrational-number weightings weren't the only game in town. Anything that introduced a similar element of randomness or unpredictability might do the trick, too. 'Having irrational numbers is only one way to get super-Turing power,' she says.

The route the trio chose was chaos. A chaotic system is one whose response is very sensitive to small changes in its initial conditions. Wire up an analogue neural net in the right way, and tiny gradations in its outputs can be used to create bigger changes at the inputs, which in turn feed back to cause bigger or smaller changes, and so on. In effect, the system becomes driven by an unpredictable, infinitely variable noise.

The researchers are working on two small prototype chaotic machines. One is a neural network based on standard electronic components, with three 'neurons' in the form of integrated circuit chips and 11 synaptic connections on a circuit board a little larger than a hardback book. The other, with 11 neurons and around 3600 synapses, uses lasers, mirrors, lenses and photon detectors to encode its information in light.

If only on a small scale that should be enough, the team thinks, to take them beyond Turing computation. It is a claim that invites plenty of scepticism. Scott Aaronson of the Massachusetts Institute of Technology voices the concern that mathematical models involving any sort of infinity always run into problems when they are forced to deal with reality. 'People ignore the fact that the physical system cannot implement the idea with perfect precision,' he says. Jérémie Cabessa of the University of Lausanne, Switzerland, who used to work with Siegelmann, is similarly doubtful about super-Turing machines in practice. 'To me, at the moment they are unbuildable,' he says. Again, it's not that the maths doesn't work – it is just a moot point whether true randomness is something we can harness, or whether it even exists. 'Does nature achieve some intrinsic randomness? If so, perhaps there really is some super-Turing ability in nature,' he says.

That question was clearly on Turing's mind: he often speculated about a connection between intrinsic randomness and the origin of creative intelligence. In 1947, he went so far as to suggest to his astounded bosses at the UK National Physical Laboratory near London that they should put radioactive radium into the Automatic Computing Engine he had devised, in the hope that its seemingly random decays would give its inputs the desired unpredictability. 'I don't think he intended to build the oracle machine,' says Siegelmann. 'What he had in mind was to build something that's more like the brain.'

Since then, building a computer with brain-like qualities has been a perennial aim, with the latest large-scale initiative being part of the Human Brain Project based at the Swiss Federal Polytechnic School in Lausanne. These endeavours, though, are all about building replica neurons with standard, digital Turing-machine technology. Younger is convinced the less-rigid approach of their chaotic neural networks is more likely to bear fruit. 'Applying this might take us towards brain-like intelligence,' he says.

## Hypercomputer hype

Younger and Redd are aware they are shooting for the moon. Even if their machine works significantly differently from a standard computer, proving that this is due to super-Turing computation will be pretty tough. At the moment, their best idea lies in a side-by-side comparison of the output of their machine and a standard computer, given the same inputs. A super-Turing machine can, in theory, produce outputs identical to those of chaotic systems, but a standard Turing machine will always start rounding them.

There has always been a lot of hype about what hypercomputers might do if they were ever to get off the ground – for example, breaking through the boundaries of conventional computation might give us a new hold on other things that currently befuddle human logic, such as quantum theory.

Most of us would be happy if an oracle would just put an end to the blue screen of death and its equivalents. While promising nothing specific of the first trials, Redd is bullish about the outcome. 'I'm actually kind of confident we'll see something significant,' he says.

## The real monsters of the deep

*They were dismissed as sailors' tall tales, writes* **Stephen Ornes**, *but they're real: huge waves that rise without warning and can destroy ships. Is there any way to predict them?*

When the cruise ship *Louis Majesty* left Barcelona in eastern Spain for Genoa in northern Italy, it was for the leisurely final leg of a hopscotching tour around the Mediterranean. But the Mediterranean had other ideas.

On 3 March 2010, storm clouds were gathering as the boat ventured eastwards out of the port at around 1 p.m. The sea swell steadily increased during the first hours of the voyage, enough to test those with less-experienced sea legs, but still nothing out of the ordinary.

At 4.20 p.m., the ship ran without warning into a wall of water 8 metres or more in height. As far as events can be reconstructed, the boat's pitch as it descended the wave's lee tilted it into a second, and possibly a third, monster wave immediately behind. Water smashed through the windows of a lounge on deck 5, almost 17 metres above the ship's water line. Two passengers were killed instantly and 14 more injured. Then, as suddenly as the waves had appeared, they were gone. The boat turned and limped back to Barcelona.

A few decades ago, rogue waves of the sort that hit the *Louis Majesty* were the stuff of salty sea dogs' legend. No more. Real-world observations, backed up by improved theory and lab experiments, leave no doubt any more that monster waves happen – and not infrequently. The question has become: can we predict when and where they will occur?

Science has been slow to catch up with rogue waves. There is not even any universally accepted definition. One with wide currency is that a rogue is at least double the significant wave height, itself defined as the average height of the tallest third of waves in any given region. What this amounts to is a little dependent on context: on a calm sea with significant waves 10 centimetres tall, a wave of 20 centimetres might be deemed a rogue.

If that seems a little lackadaisical, for a long time the models oceanographers used to predict wave heights suggested anomalously tall waves barely existed. These models rested on the principle of linear superposition: that when two trains of waves meet, the heights of the peaks and troughs at each point simply sum. It was only in the late 1960s that Thomas Brooke Benjamin and J. E. Feir of the University of Cambridge spotted an instability in the underlying mathematics. When longer-wavelength waves catch up with shorter-wavelength ones, all the energy of a wave train can become abruptly concentrated in a few monster waves – or just one.

Longer waves travel faster in the deep ocean, so this is a perfectly plausible real-world scenario. The pair went on to test the theory in a then state-of-the-art 400-metre-long towing tank, complete with wave-maker, at the UK National Physical Laboratory facility on the outskirts of London. Near the wave-maker, which perturbed the water at varying speeds, the waves were uniform and civil. But about 60 metres on they became distorted, forming into short-lived, larger waves that we would now call rogues (though to avoid unwarranted splashing, the initial waves were just a few centimetres tall).

It took a while for this new intelligence to trickle through. 'Waves become unstable and can concentrate energy on their own,' says Takuji Waseda, an oceanographer at the University of Tokyo in Japan. 'But for a long time, people thought this was a theoretical thing that does not exist in the real oceans.'

Theory and observation finally crashed together in 1995 in the North Sea, about 150 kilometres off the coast of Norway. New Year's Day that year was tumultuous around the Draupner sea platform, with a significant wave height of 12 metres. At around 3.20 p.m., however, accelerometers and strain sensors mounted on the platform registered a single wave towering 26 metres over its surrounding troughs. According to the prevailing wisdom, this was a once-in-10,000-year occurrence.

The Draupner wave ushered in a new era of rogue-wave science, says physicist Ira Didenkulova at Tallinn University of Technology in Estonia. In 2000, the European Union initiated the three-year MaxWave project. During a three-week stretch early in 2003, it used boat-based radar and satellite data to scan the world's oceans for giant waves, turning up 10 that were 25 metres or more tall.

We now know that rogue waves can arise in every ocean. The North Atlantic, the Drake Passage between Antarctica and the southern tip of South America, and the waters off the southern coast of South Africa are particularly prone. Rogues possibly also occur in some large freshwater bodies such as the Great Lakes of North America. That casts historical accounts in a new light, and rogue waves are thought to have had a part in the unexplained losses of some 200 cargo vessels in the two decades preceding 2004. More recently, what is thought to have been a freak wave struck the cruise ship *Marco Polo* in the English Channel in 2014, smashing windows in a restaurant on deck 6 and killing a passenger.

## Rogue elements

So rogue waves exist, but what makes one in the real world?

Miguel Onorato at the University of Torino, Italy, has spent more than a decade trying to answer that question. His tool is the non-linear Schrödinger equation, which has long been used to second-guess unpredictable situations in both classical and quantum physics. Onorato uses it to build computer simulations and guide wave-tank experiments in an attempt to coax rogues from ripples.

Gradually, Onorato and others are building up a catalogue of real-world rogue-generating situations. One is when a storm swell runs into a powerful current going the other way. This is often the case along the North Atlantic's Gulf Stream, or where sea swells run counter to the Agulhas current off South Africa. Another is a 'crossing sea', in which two wave systems – often one generated by local winds and a sea swell from further afield – converge from different directions and create instabilities.

Crossing seas have long been a suspect. A 2005 analysis used data from the maritime information service Lloyd's List Intelligence to show that, depending on the precise definition, up to half of ship accidents chalked up to bad weather occur in crossing seas.

In 2011, the finger was pointed at a crossing sea in the Draupner incident, and Onorato thinks it might also have been the *Louis Majesty*'s downfall. When he and his team fed wind and wave data into his model to 'hindcast' the state of the sea in the area at the time, it indicated that two wave trains were converging on the ship, one from a north-easterly direction and one more from the south-east, separated by an angle of between 40 and 60 degrees.

Simpler situations might generate rogues, too. In 2013,

Waseda revisited an incident in December 1980 when a cargo carrier loaded with coal lost its entire bow to a monster wave with an estimated height of 20 metres in the 'Dragon's Triangle', a region of the Pacific south of Japan notorious for accidents. A Japanese government investigation had blamed a crossing sea, but when Waseda used a more sophisticated wave model to hindcast the conditions, he found it likely that a strong gale had poured energy into a single wave system far larger than conventional models allowed.

He thinks such single-system rogues could account for other accidents, too – and that the models need further updating. 'We used to think ocean waves could be described simply, but it turns out they're changing at the same pace and same time scale as the wind, which changes rapidly,' he says. In 2012, Onorato and others showed that the models even allow for the possibility of 'super rogues' towering as much as 11 times the height of the surrounding seas, a possibility since borne out in water-tank experiments.

## Early warning

With climate change potentially whipping up more intense storms, such theoretical possibilities are becoming a serious practical concern. From 2009 to 2013, the EU funded a project called Extreme Seas, which brought shipbuilders together with academic researchers including Onorato, with the aim of producing boats with hulls designed to better withstand rogue waves.

That is a high-cost, long-term solution, however. The best defence remains simply knowing when a rogue wave is likely to strike. 'We can at least warn that sea states are rapidly changing, possibly in a dangerous direction,' says Waseda.

Various indices have been developed that aim to convert

raw satellite and sea-state data into this sort of warning. One of the most widely used is the Benjamin–Feir index, named after the two pioneers of rogue-wave research. Formulated in 2003 by Peter Janssen of the European Centre for Medium-Range Weather Forecasts in Reading, UK, it is calculated for sea squares 20 kilometres by 20 kilometres, and is now incorporated into the centre's twice-daily sea forecasts. 'Ship routing officers use it as an indicator to see whether they should go through a particular area,' says Janssen.

The ultimate aim would be to allow ships to do that themselves. Most large ocean-going ships now carry wide-sweeping sensors that determine the heights of waves by analysing radar echoes. Computer software can turn those radar measurements into a three-dimensional map of the sea state, showing the size and motions of the surrounding swell. It would be a relatively small step to include software algorithms that can flag up indicators of a sea about to go rogue, such as quickly changing winds or crossing seas. Such a system might let crew and passengers avoid at-risk areas of a ship.

The main bar to that happening is computing power: existing models can't quite crunch through all the fast-moving fluctuations of the ocean rapidly enough to generate fine-grained warnings in real time. For Waseda, the answer is to develop a central early warning system, such as those that operate for tsunamis and tropical storms, to inform ships about to leave port. Thanks to our advances in understanding a phenomenon whose existence was doubted only decades ago, there is no reason now why we can't do that for rogue waves, says Waseda. 'At this point it's not a shortage of theory, but a shortage of communication.'

## Five mythical animals that turned out to be real

*Since we began exploring the world, travellers have returned with stories of distant lands filled with improbable creatures.* **Michael Marshall** *chronicles five that turned out to be true.*

In 2017, biologists in Taiwan discussed whether or not to re-introduce the Formosan clouded leopard, a creature so mysterious that some have claimed it may never have existed. It's not an entirely unusual state of affairs. Explorers have claimed to have seen bizarre animals over the centuries, only to be exposed as hoaxers. But not always. Sometimes the most outlandish creatures turn out to be extremely real.

### Duck-billed platypus

It is perhaps no surprise that the platypus was once thought to be a hoax. It looks a bit like a mole but has a duck's bill. Not only did this strange-looking mix of mammal and bird not fit with what was then known of biology, it was also immediately obvious how the hoax might have been achieved, with little more than scissors, thread and a sewing needle.

The platypus was scientifically described for the first time in 1799 by the British naturalist George Shaw, based on a skin sent by John Hunter, then the governor of Australia. Shaw admitted to being suspicious: 'it naturally excites the idea of some deceptive preparation by artificial means,' he wrote.

But Shaw couldn't find any telltale stitching. Over the years more specimens followed, as well as descriptions of the animals in the wild. By 1823, the anatomist (and grave robber) Robert Knox could write that, while the extraordinary nature of the platypus had been 'sufficient to rouse the suspicions of the scientific naturalist', nevertheless 'these conjectures were immediately dispelled by an appeal to anatomy'.

Since then, the story of the platypus has only grown stranger. Its genome is deeply peculiar, it lays eggs much like birds do, has venomous spurs on its hind legs, and it is descended from 'King Kong platypuses' which were a metre long.

## Okapi

For centuries, European travellers in West Africa – particularly in what is now the Democratic Republic of the Congo – reported seeing glimpses of a mysterious animal in the forest.

The descriptions were sketchy. It was hoofed, perhaps looked a little like a deer, but had stripes on its rear end that suggested it might be a forest-dwelling zebra. Nobody could catch one or even get a good look at one. For want of a better name, people began referring to it as the 'African unicorn'.

From 1871 onwards, the British explorer Henry Morton Stanley undertook several expeditions to Africa. In his 1890 book *In Darkest Africa*, he mentioned that a group called the Wambutti knew of 'a donkey' called the 'atti'. There the matter rested until 1901, when explorer and colonial administrator Harry Johnston mounted a determined search. Local people had told him about a forest animal called the 'o'api' – the name Stanley had evidently misheard. He managed to obtain some skins to send to London, where they caused a great deal of confusion and were briefly misidentified as zebra skins.

Eventually, it was recognised that the okapi belonged to a new genus, and it was given the name *Okapia johnstoni*. It remains incredibly elusive. Very few photographs exist of them in the wild. It is a threatened species, due to illegal logging and the continuing unrest in the DRC. Unexpectedly, its closest living relatives are giraffes.

## Giraffe

Speaking of giraffes, there is a much-repeated canard that when Europeans first found out about giraffes, they thought they were a cross between a camel and a leopard, not a species in their own right. They didn't, but there is a tale of confusion here nonetheless.

The scientific name of the giraffe is *Giraffa camelopardalis*. The latter part of the name dates back to the Roman Empire, when Julius Caesar brought a giraffe back to Rome from Alexandria: the first time in recorded history that a giraffe visited Europe. Various writers described the animal in terms of camels and leopards. The Roman senator and historian Cassius Dio, in his *Roman History*, described 'the so-called camelopard'.

But it seems Dio didn't think the giraffe was any kind of cross. Instead, he was just helping his readers picture what it looked like using animals they would have known. He describes the animal as 'like a camel in all respects' except for its unusual height and proportions, and notes that 'its skin is spotted like a leopard, and for this reason it bears the joint name of both animals'.

Camelopard, then, is not so much a misidentification as a neologism.

Modern science has revealed that giraffes hum and that baby giraffes are inveterate milk thieves. It has also found answers to the long-standing mystery of how giraffes got such long necks.

## Colossal squid

For centuries, sailors told tales of enormous tentacled creatures that could drag entire ships down to Davy Jones' Locker.
The Scandinavian legend of the kraken is just one example.

Such things remained mythical until 1925, when G. C.
Robson published a description of a squid called *Mesony-
choteuthis hamiltoni*. Robson based his description on two
tentacles found in the stomach of a sperm whale. Only a few
specimens have been found in the intervening 90 years, so
our knowledge of this squid is still sketchy.

What is clear is that *M. hamiltoni* is the largest known
species of squid, with at least one specimen measured at 4.5
metres long. It has been given the moniker 'colossal squid',
not to be confused with the jumbo squid and giant squid,
which are different species.

However, the idea that it could sink a ship or pose any
kind of threat to humans on the surface appears to be a
fantasy. Colossal squid live deep underwater where the pressure is intense, and they have adapted accordingly. If they
find themselves in surface waters their bodies become floppy
and incapable.

## Narwhal

Thanks to their huge tusks, narwhals are often called 'unicorns
of the sea' and for a long time people seem to have thought
that was literally true. The earliest attempt at a scientific
description may have been that by Nicolaes Tulp, a doctor
and anatomist working in Amsterdam in the 1600s, who was
famously painted by Rembrandt.

Tulp wrote a monumental medical book called *Observationes
Medicae*, in which he included a few snippets of natural

history. The book contains what may be the first Western illustration of an orangutan, and a drawing of the horn of a 'unicornum marinum', or 'marine unicorn'. It is almost certainly a narwhal tusk. These were often presented as belonging to unicorns, and were popular in cabinets of curiosities.

It was not until 1758, more than 100 years after Tulp's description, that Linnaeus described narwhals for the first time in the tenth edition of his *Systema Naturae*. He correctly identified them as related to whales. Today narwhals are classed as 'near threatened' because of continued hunting and the risk that climate change will melt the pack ice where they live.

## Bonus beast: King Louie

Disney gave King Louie an upgrade it its 2016 remake of *The Jungle Book* when it reimagined the king of the jungle as an impossibly large orangutan. In fact, the creature was closer to reality than might first appear.

Once upon a time, the forests of East Asia were home to the largest of all apes. *Gigantopithecus* reached 3.5 metres tall and weighed 540 kilograms. At that size, it wasn't quite as large as King Kong, but would have looked down on Chewbacca from on high.

# 2 You Are Not Who You Think You Are

The bad news first: almost everything you think you know about yourself is wrong. You're not really human, you're not rational, and you might not even be real. But the good news is that you are more amazing than you ever realised. You can outrun horses, read minds and intuit complex physics. Your age really is just a number, and you can train your mind to make you braver, kinder and happier. This chapter deals with the most mind-blowing creature on Earth: you.

# You get mutant powers from outsider genes

*Genes from other species, and cells from your relatives, live inside your body, writes **Sean O'Neill** – and they hint at how we can improve ourselves.*

Let's begin with the obvious. You are the product of billions of years of evolution, the accumulation of trillions of gene-copying errors. That's what led single cells to evolve into jellyfish, ferns, warthogs and humans. Without mutations, life would never have evolved into Darwin's 'endless forms most beautiful', and you would never have seen the light of day.

Today, while most of our genes are undeniably *Homo sapiens*, many of us also carry DNA from other species. We have known for a decade that people of non-African descent inherit between 2 and 4 per cent of their DNA from Neanderthals. And we now know that DNA from several other extinct human species is also still in circulation, on every continent including Africa.

Not only do you carry DNA from other species, you probably also play host to other people's cells. Before you were born, your mother's cells crossed the placenta into your bloodstream. Decades later, some of these migrants are still there in your blood, heart, skin and other tissues. This 'microchimeric' exchange was mutual: if you are a mother, your children may still be inside you in the form of their embryonic stem cells.

You may even be carrying cells from your grandmother and any older siblings. Because microchimeric cells persist for a long time, there is a chance that during pregnancy your mother was still carrying cells from any previous children she had, as well as cells from her own mother – and she may have shared some with you.

Maternal microchimerism is extensive, says Lee Nelson at the University of Washington in Seattle, and probably useful too. 'There are so many examples in biology where organisms thrive as a result of exchange – why wouldn't it also be useful for humans to exchange cellular material?' Foetal cells may help to repair a mother's damaged heart tissue and lower her risk of cancer. Other research shows that mothers can end up with their child's DNA in their brains, something that may even be linked to a reduced risk of the mother developing Alzheimer's.

In future, we could become mutants by design. Gene-editing tools like CRISPR should allow genetic diseases to be treated by injecting genes into the body. For example, a small number of people with a mutation in the $CCR5$ gene, which supplies a protein to the surface of white blood cells, are resistant to HIV. CRISPR opens the possibility of inserting that mutation into the DNA of others, giving them a genetic vaccine against the virus.

From there, it's only a baby-step to genetic superpowers. Ethical questions notwithstanding, future generations could be enhanced with genes for extra-strong bones, lean muscles and a lower risk of cardiovascular disease and cancer. A mutation in the $ABCC11$ gene currently found in about 1 in 50 Europeans even renders underarms odourless. Think of the savings on deodorant. Be warned, however: this mutation also makes your ear wax dry up. Swings and roundabouts.

## Your body is a nation of trillions

*Think you're only human? Legions of creatures inhabit the cracks, contours and crevices of your body – and they all contribute to who you are, says* **Daniel Cossins**.

Last night, while you were sleeping, legions of eight-legged creatures had an orgy between your eyebrows. No, you haven't suddenly been invaded by sex tourists. *Demodex* mites, close relatives of ticks and spiders, are permanent and mostly harmless residents of the human face.

'Every person we've looked at, we've found evidence of face mites,' says Megan Thoemmes at North Carolina State University in Raleigh. 'You can have thousands living on you and never even know they're there.'

Growing up to 0.4 millimetres long, these beasts spend their days buried head-down in hair follicles gorging on who-knows-what and crawling out under cover of darkness to copulate. They have no anus, so on death disgorge a life-time of faeces into your pores.

Before you lunge for the exfoliating brush: *Demodex* mites are far from your only microscopic residents. You host aston-ishing biodiversity, from anus-less arthropods to pubic lice to all manner of bacteria and fungi, and without it you wouldn't be who you are. 'Each of us is really a complex consortium of different organisms, one of which is human,' says Justin Sonnenburg at Stanford University in California.

Our resident aliens aren't all benign. There are big beasts like parasitic worms: roundworm, hookworm and whipworm are prevalent in the developing world, and pinworm still infects kids in the West. Then there are hidden viruses such as *Herpes simplex*, which lies dormant inside the nerve cells of two-thirds of people until it mistakes your sniffles for a deadly fever and attempts to save itself by rushing outwards, causing cold sores.

By far the dominant group, however, are bacteria. You have at least as many bacterial cells as human cells, perhaps 10 times more. Only recently have we begun to grasp the extent of their diversity, and there's plenty left to discover. We've even found bacteria that survive by parasitising other bacteria. They live in your spit.

Similar battles play out across your many habitats, from the caves of your nostrils and your anal-genital badlands to the crevices between your toes where the fungus *Trichophyton rubrum* can flare up as athlete's foot. All of these critters are constantly shedding from your skin and lungs, forming your own unique cloud of airborne bacteria that follows you everywhere.

But the densest microbial gathering is in our gut, a community that affects aspects of health from digestion and immune defences to possibly even mood and behaviour. In mice, seeding the gut with *Lactobacillus rhamnosus* bacteria has been shown to alleviate anxiety, perhaps by producing molecules that alter brain chemistry.

The balance of gut microbiota can shift rapidly in response to diet and lifestyle. To tend it you need to feed it right. Your best bet isn't much-hyped probiotics or live bacteria, but simply to eat more fibre, the preferred meal for a group of bacteria with potent anti-inflammatory powers. 'It has been known for a long time that plant-based fibre is associated with good health,' says Sonnenburg. 'Now we know why.'

# There is a physics genius inside your brain

*Without even realising, you perform fiendishly complex real-time calculations and predict the future like no other species can, says **Richard Webb**.*

The washing-up pile wobbles precariously as you balance another saucepan at its summit. For a second, it looks like the whole stack will come down. But it doesn't. Swiftly, instinctively, you save it.

Congratulations – not just on another domestic disaster averted, but also on showing a peculiarly human genius. Octopuses rival our dexterity, New Caledonian crows have a frighteningly clever way with tools and chimps beat us in tests of short-term memory. But no other species can perform complex, real-time calculations of their physical environment and generate specific, actionable predictions quite like the ones that rescued your crockery. 'It's kind of amazing,' says artificial intelligence researcher Peter Battaglia from Google DeepMind in London. 'To me it defies my ability to understand.'

In 2013, Battaglia and two colleagues showed that our inbuilt 'physics engine' works in a similar way to a graphics engine, software used in video games to generate a realistic playing environment. It is programmed with rules about objects' physical behaviour, and uses limited real-time inputs (from a player in a game, from our senses in reality) plus probabilistic inference to generate a picture of what comes next. 'What you have in your head is some means for running a simulation,' says Battaglia. 'You make a 3D model of what's around you and press the run button, it tells you what will happen. It's a way to predict the future.'

In 2016, Jason Fischer at Johns Hopkins University in

Baltimore, Maryland, and his colleagues scanned the brains of people doing a task involving physics intuition – predicting how a tower of stacked wooden blocks would fall – and showed that the physics engine sits in specific brain regions. Areas of the motor cortex associated with the initiation of bodily movement consistently lit up during the first task, but not on a second, purely mathematical task, estimating the number of different coloured blocks in the tower.

That was surprising at first, says Fischer. 'But on the other hand it makes perfect sense: you don't execute any action without mental models.' So our inbuilt genius won't necessarily help us with physics as an academic discipline, which relies on different brain circuits. That much is clear in experiments where researchers get people to draw the predicted path of a falling object, says Fischer: their intuitions are completely off. But have them catch the same falling object, forcing them to engage their motor system, and they're spot on.

There's still a lot to learn about how we generate our simulations – not least given that our device's power consumption, at around 20 watts, is less than a tenth that of a medium-range graphics card. 'The type of processing we use is clearly vastly more efficient,' says Battaglia.

But we should be aware of our limitations, too. Our physics engine is programmed with the equations of classical mechanics, which describe the visible world around us – things like falling plates. It does not work so well on less obvious layers of reality. 'Understanding electromagnetism and quantum mechanics, our instincts are not going to be so useful,' says Fischer. There, things don't stack up so easily.

## You are frighteningly easy to manipulate

*Your behaviour is heavily influenced by your environment and the people around you – but **Julia Brown** finds there are easy ways you can take back control.*

You wouldn't stand facing the back of the lift or sit out front in the garden, would you? Well there you are. Proof positive that you're not in control of your actions – the people around you are. And not just them. Your environment controls you, as do habits you don't even know you have. But realise what's really pulling your strings, and you can work out how to manipulate yourself for the better.

US social scientist Roger Barker was the first to notice this sort of environmental control. Back in the 1950s, he observed the population of a small US town, and realised that the best predictor of a person's behaviour was not personality or individual preferences, but their surroundings. People in a shop behaved as people in shops do. Ditto for libraries, churches, bars, music classes, everything.

More recently, Wendy Wood of the University of Southern California and her colleagues have shown how almost half of the behaviours we adopt in any given situation are habitual – an automated action learned by repetition until we do it without thinking. 'These were a wide range of behaviours,' says Wood, 'including eating, napping, watching TV, exercising and talking with others.'

Social control bubbles up from beneath, too. 'Reputations are so important in the social world,' says Val Curtis, who studies behaviour change at the London School of Hygiene and Tropical Medicine. Society is founded on cooperation, and we can't benefit from it unless we gain acceptance by adhering to the unwritten rules. So we face front in the lift.

We work this way because neurons are expensive to run. If we had to do everything consciously, we would have no energy for anything else – automation frees up processing power. We notice that when we lose the comfort blanket of subconscious control. 'If you have ever walked into a restaurant in a foreign country, you are almost paralysed until you work out what everyone else is doing and then copy them,' says Curtis.

Identifying your unconscious workings provides you with ways to fine-tune your behaviour. For a start, if you want to change bad habits, have a look at where and how you enact them, and then try to disrupt that pattern. If you want to stop smoking, avoid the places where you are likely to spark up, or move your cigarettes out of sight. If you want to start eating more healthily, stop meeting friends for lunch at a burger restaurant. 'Yes, you think now that you'll order the salad, but when you get there, the cues and smells will be hard to resist,' says Wood.

Curtis has used such insights to develop ways to encourage handwashing with soap in India and to modify the tendency for mothers in Indonesia to feed their children unhealthy snacks. She suggests we can all prime ourselves in similar ways. If you think you ought to do some exercise but don't really feel like it, just put your running gear on anyway, and wait and see what happens, she says. 'The kit takes you for a run. You let it control your behaviour.'

## You are a fantasist

*Think you're saner, smarter and better-looking than the average? Well so does everyone else. Recognising our* *delusions is the first step to doing better, writes **Tiffany O'Callaghan**.*

Ever had the sense that everyone else is an idiot? Maybe that's a tad overblown, but when it comes to smarts, looks, charisma and general psychological adjustment, there's no denying you are a cut above the average person in the street. Or on the road: have you seen how those jerks drive?

Well, here's the bad news. Pretty much everyone else is thinking the same thing.

The phenomenon of self-enhancement – viewing ourselves as above average – applies across human ages, professions and cultures, and to capabilities from driving to playing chess. It does have advantages. People who are more impressed with themselves tend to make better first impressions, be generally happier and may even be more resilient in the face of trauma. High self-estimation might also let you get ahead by deceiving others: anthropologist Robert Trivers at Rutgers University in New Brunswick, New Jersey, argues that when we've tricked ourselves, we don't have to work so hard to trick others, too.

Confidence also helps in finding a romantic partner, and so in reproduction. When it comes to overestimating our looks, we're all at it – although men are on average worse offenders than women. According to a 2016 study by Marcel Yoder of the University of Illinois in Springfield and his colleagues, men seem to suffer from a 'frog prince' delusion: they accurately assess other people's lesser perception of them, while persisting in a more positive perception of themselves.

The real downsides come when you're less aware of how others perceive you. If you are self-confident without being self-aware, you are likely to be seen as a jerk. 'It's hard to come off as humble or modest when you're clueless about how other people see you,' says Yoder. Plus we may make bad decisions on the basis of an inflated sense of expertise or understanding.

Particularly in the political arena, our 'bias blind spot' – a belief that our world view is based on objective truth, while everyone else is a deluded fool – can become problematic, especially as the echo chamber of social media exposes us to fewer contrary views. 'It can make opposing parties feel that the other side is too irrational to be reasoned with,' says Wenjie Yan, who studies communication at Washington State University in Pullman.

So how can we preserve the good while avoiding the downsides? Different strategies and training programmes do exist for overcoming our inbuilt biases. Most begin by simply making people aware of them and how they can affect our decision-making.

At home, we can use an exercise that psychologists call 'perspective-taking'. This amounts to trying to see a dispute from the other person's point of view, says Irene Scopelliti, who studies decision-making at City University of London. She also points out that acting when you're all riled up – in a state of high emotion – only entrenches your bias. 'We know how to make unbiased decisions, but often emotion pushes us, or we aren't willing to put in the effort,' she says. But then comes the good news: 'practice can make us better.'

## Evolution made you a scaredy-cat

*An inbuilt fear factory makes us err on the side of caution. But by engaging a different way of thinking we can stop panicking and weigh up the real risks, says* **Sally Adee**.

In the aftermath of 11 September 2001, most people in the US believed that they or their families were highly likely to become victims of terrorist attacks. 'Which is just off the charts crazy when you think about it for even a minute,' says Dan Gardner, an author and risk consultant based in Canada. Instead of boarding planes, people in the US got in their cars. Over the following five years, the annual death toll on the road was on average 1100 higher than it had been in the five preceding years.

We are, in general, appalling at assessing risk: driving is inherently riskier than flying, terrorists or no terrorists. We also underestimate our chance of divorce, and spend more than is rational on lottery tickets and less than is rational on climate change. We fear our kids being abducted, so drive them to school, ignoring the greater risks that poses to their health and well-being.

How to do better? First, switch off your gut. Psychologists characterise our risk problem as a clash between system 1 and system 2 thinking. System 1 is the product of evolved biases shaped over thousands of years. 'If you saw a shadow in the grass and it was a lion and you lived to tell the tale, you'd make sure to run the next time you saw a shadow in the grass,' says Gardner.

This inbuilt fear factory is highly susceptible to immediate experience, vivid images and personal stories. Security companies, political campaigns, tabloid newspapers and ad agencies prey on it. System 1 is good at catastrophic risk, but less good at risks that build up slowly over time – hence our lassitude in the face of climate change or our expanding waistlines.

So when your risk judgement is motivated by fear, stop and think: what other, less obvious risks might I be missing? This amounts to engaging the more rigorous, analytical system 2. People who deal with probability and risk professionally have been found to use system 2 more, among them bookies, professional card players – and weather forecasters.

'Meteorologists get a bad rap,' says Gardner, 'but they tend to be highly calibrated, unlike most of us.'

These people receive precise, near-immediate feedback about their predictions – abuse for a false weather forecast, or a crucial card trick lost – which helps them constantly recalibrate their risk thermometer. That's something we can all do. 'Choose something specific you want to improve your risk intelligence for,' says Dylan Evans, a risk researcher. 'What time will your spouse be home tonight? Make bets with yourself. Were you right? Keep track.'

That sounds trivial in the home, but it's crucial in business. Part of the problem in the run-up to the financial collapse of 2008 was that individuals were no longer accountable for their own actions, says Andre Spicer, who studies organisational behaviour at City University of London. 'At banks, there was no direct relationship between what you did and the outcome,' says Spicer. 'That produced irrational decisions.'

There's one feature you see over and over in people with good risk intelligence, says Gardner. 'I think it wouldn't be too grandiose to call it the universal trait of risk intelligence – humility.' The world is complex – be humble about what you know, and you'll come out better.

## Mindfulness can defeat your inner bigot

*We are wired to be prejudiced and a bit racist, says* **Caroline Williams**, *but our instinct for collaboration can trump our worst instincts.*

From Brexit to President Trump, recent political events have let some nasty cats out of the bag. Racists and xenophobes are on the march. But perhaps that shouldn't be so surprising: after all, that is what we are.

Here's the unpalatable truth: we are biased, prejudiced and quite possibly a little bit racist. Psychologists have long known that we put people into little mental boxes marked 'us' and 'them'. We implicitly like, respect and trust people who are the most similar to us, and feel uncomfortable around everybody else. And before you deny it, this tendency towards in-group favouritism is so ingrained we often don't realise we are doing it. It is an evolutionary hangover affecting how the human brain responds to people it perceives as different.

In one study from 2000, just showing participants brief flashes of faces of people of a different race was enough to activate the amygdala, part of the brain's fear circuitry, even though the participants felt no conscious fear. According to more recent research, however, the amygdala doesn't just control fear; it responds to many things, calling on other brain areas to pay attention. So although we're not automatically scared of people different from us, we are hardwired to flag them. Evolutionarily, that makes sense: it paid to notice when someone from another tribe dropped by.

We're also prone to dehumanisation. When Susan Fiske at Princeton University scanned volunteers' brains as they looked at pictures of homeless people, she found that the medial prefrontal cortex, which is activated when we think about other people, stayed quiet. Volunteers seemed to be processing the homeless people as subhuman.

'The bad news is how fast this automated "us" and "them" response is, and how wired-in it is,' says Fiske. 'The good news is that it can be overcome depending on context.' In both the homeless study and a rerun of the amygdala study, Fiske found that fear or indifference quickly disappeared when participants were asked questions about what kind of food the other person might enjoy. 'As soon as you have a basis for dealing with a person as an individual, the effect is not there,' says Fiske.

What's more, what we put in the 'them' and 'us' boxes is remarkably flexible. When Jay Van Bavel at New York University created in-groups including people from various races, volunteers still preferred people in their own group, regardless of race. All you have to do to head off prejudice, it seems, is to convince people they are on the same team.

We are also instinctively cooperative, at least when we don't have time to think about it. Yale University psychologist David Rand asked volunteers to play gambling games in which they could choose to be selfish, or cooperate with other players for a slightly lower, but shared, payoff. When pressed to make a snap decision, people were much more likely to cooperate than when given time to mull it over.

So perhaps you're not an arsehole after all – if you know when to stop to think about it and when to go with your gut. Maybe, just maybe, there is hope for the world.

## We're all reading each other's minds all the time

*Your power to predict what other people think is the secret sauce of culture and social connections. And there's scope for us all to improve, writes* **Gilead Amit**.

Meet Sally and her flatmate Andy. Sally has made a birthday cake for Andy, and leaves it in the fridge while she pops out to buy some candles. While she's gone, Andy sneaks into the kitchen, takes the cake and hides it on a shelf to consume at leisure. When Sally comes back, where does she think the cake will be?

If you answered 'the fridge' then congratulations: you understand that, based on what they know, people can have different views from you. You possess a 'theory of mind' –

something that informs your every waking moment, says Josep Call, a psychologist at the University of St Andrews. 'When we get dressed in the morning, we're constantly thinking about what other people think about us.' No other animal can match our ability, making it the essential lubricant for the social interactions that set humans apart.

Take the arts. Artists must be able to imagine what their audiences will think of their characters. Without a theory of mind, there would be no compelling TV soaps, sculptures or books. Some think William Shakespeare must have had a particularly well-developed theory of mind to create such rich, complex characters.

Mind reading is also crucial for societal norms. 'People not only respond to what you do, but to what you intend to do,' says Call. If you hit someone with your car, the difference between a verdict of murder or manslaughter depends on your intent.

Yet we can't all read minds equally well, says Rory Devine, a psychologist at the University of Cambridge. Most of us come a cropper when attempting nested levels of mind reading. Think of Sally hunting for her cake again, but imagine where she might look if we take into account what she thinks about how Andy's mind works. The more recursive steps we add, the more we stumble. 'When you go beyond five levels, people get really, really bad,' says Call.

Being a good mind reader pays. Children who are relatively proficient later report being less lonely, and their teachers rate them as more sociable.

We may be able to improve our skills. We know our mind reading apparatus mostly develops before the age of 5, and the principal factor that determines its development is whether our families and friends talk much about the emotions and motivations of others. 'The ability to read minds is something we might learn gradually from the guidance of others,' says Devine.

This suggests that it could help to just think about what it's like to be in other people's shoes. In 2014, Devine and his colleagues showed that this learning can continue far beyond early childhood. When they asked 9- and 10-year-old children to read and discuss short vignettes about social situations, the team found they developed better mind-reading skills than children in a control group. Similar improvements have also been seen in people over the age of 60. You're never too old to be a better mind reader.

## You are the greatest runner on Earth

*Other species might be better at speed or distance, but no species can run faster, further under all conditions than humans can. And yes, says **Catherine de Lange**, that includes you.*

In October 2016, Daniel Lieberman set out on the race of a lifetime. A 25-mile slog in the Arizona heat, climbing a mountain more than 2,000 metres tall. To top it all, 53 of his competitors had four legs. This was the 33rd annual Man Against Horse Race. Lieberman, by his own admission not a great runner, outran all but 13 horses – and so could you.

Lieberman studies human evolutionary biology at Harvard University, and part of his work over the past 15 years has focused on a unique set of adaptations that suggest modern humans evolved not just to walk, but to run long distances.

One is our cooling equipment. 'The fact we have sweat glands all over our body and we've lost our fur enables us to dump heat extremely effectively,' says Lieberman. This is crucial when running for long periods. It helps to explain why animals struggle to beat us in the heat, even though sled dogs can run more than 100 kilometres a day pulling humans in cold climates. Hence also Lieberman's success in Arizona. 'The hotter it is, the better humans are able to run compared with horses,' he says.

Then there are adaptations that offset our clumsy, inefficient bipedal frames. Short toes and large gluteal muscles assist with balance and stability. The Achilles tendon and other springs in the feet and legs help us to store and release energy. We tend to have a high proportion of slow-twitch muscle fibres, which produce less power but take longer to tire than the short-burst, fast-twitch fibres needed for sprinting.

The nuchal ligament at the base of the skull also helps to keep our heads, and therefore our gaze, steady when we run. Other decent runners such as dogs and horses have one, but they're not found in poor runners such as pigs and non-human primates or early hominids like *Australopithecus*. Many of these adaptations are specific to running, suggesting we're not just good at it because we are good walkers.

One theory is that we began running as scavengers, where an ability to outrun other carnivores to reach fresh meat was to our advantage. As we improved, we became better hunters, able to track and outrun our prey over large distances before we had spears and arrows. This all helped to provide us with the extra protein we needed to acquire our greatest advantage: a bigger brain. 'The features that we see in the fossil record that are involved in running appear about when we start to see evidence for hunting. And soon thereafter their brains start to get bigger,' says Lieberman.

So can you unleash your inner marathon runner? In a word, yes. Genetics is important but training is key, says sports scientist Chris Easton at the University of the West of Scotland. You'll need stronger leg and bum muscles, to be sure, but you can get these simply by starting to run. You will find it hard to increase the proportion of slow-twitch muscle fibres you have, but if you find yourself flagging, take your time and take comfort in the fact we evolved to jog, rather than sprint, over the finish line. 'Millions of people run marathons and people tell us we are crazy,' says Lieberman. 'Actually, it's part of who we are.'

## Think you're an atheist? Heaven forfend!

**Graham Lawton** reveals that your default is to believe in the supernatural, and there is no manual override.

Fingers crossed, touch wood. By the time you finish this, you'll believe you believe in the supernatural.

For most of us, that is a given. The vast majority of people are religious, which generally entails belief in a supernatural entity or three. And yet amid the oceans of religiosity are archipelagos of non-belief. Accurate numbers are hard to come by, but even conservative estimates suggest that half a billion people around the world (and counting) are non-religious.

But are they, really? Among the scientists who study the cognitive foundations of religious belief, there is a widespread consensus that atheism is only skin-deep. Scratch the surface of a non-believer and you'll find a writhing nest of superstition and quasi-religion.

That's because evolution has endowed us with cognitive tendencies that, while useful for survival, also make us very receptive to religious concepts. 'There are some core intuitions that make supernatural belief easy for our brains,' says psychologist Ara Norenzayan at the University of British Columbia.

One is the suite of cognitive abilities known as theory of mind, which enables us to think about and intuit other people's thoughts. That's damn useful for a social species like us, but also tricks us into believing in disembodied minds with mental states of their own. The idea that mind and body are distinct entities also seems to come instinctively to us. Throw in teleology – the tendency to seek cause and effect everywhere, and see purpose where there is none – and you can see why the human brain is a sitting duck.

The same thought processes probably underlie belief in other supernatural phenomena such as ghosts, spiritual healing, reincarnation, telepathy, astrology, lucky numbers and Ouija boards. These are almost as common as official religious beliefs; three-quarters of Americans admit to holding at least one of 10 common supernatural beliefs.

With all this supernatural equipment filling our heads, atheism and scientific materialism are hard work. Overriding inbuilt thought patterns requires deliberate and constant effort, plus a learned reference guide to what is factually correct and what is right and wrong. Just like a dieter tempted by a doughnut, willpower often fails us.

Many experiments have shown that supernatural thoughts are easy to invoke even in people who consider themselves sceptics. Asked if a man who dies instantly in a car crash is aware of his own death, large numbers instinctively answer 'yes'. Similarly, people who experience setbacks in their lives routinely invoke fate, and uncanny experiences are widely attributed to paranormal phenomena.

Obviously, it is impossible to prove that everyone falls prey to supernatural instincts. 'There is no more evidence than a few studies, and even they do not provide enough support for the argument,' says Marjaana Lindeman, who studies belief in the supernatural at the University of

Helsinki. Nonetheless, the supernatural exerts a pull on us that is hard to resist. If you're still under the illusion that you are a rational creature, that really is wishful thinking.

## You might be a hologram

*You, I and the entire universe may be a hologram – and a major new experiment is dedicated to finding out, says* **Marcus Chown**.

Take a look around you. The walls, the chair you're sitting in, your own body – they all seem real and solid. Yet there is a possibility that everything we see in the universe – including you and me – may be nothing more than a hologram.

It sounds preposterous, yet there is already some evidence that it may be true. If it does turn out to be the case, it would turn our common-sense conception of reality inside out.

The idea has a long history, stemming from an apparent paradox posed by Stephen Hawking's work in the 1970s. He discovered that black holes slowly radiate their mass away. This Hawking radiation appears to carry no information, however, raising the question of what happens to the information that described the original star once the black hole evaporates. It is a cornerstone of physics that information cannot be destroyed.

In 1972, Jacob Bekenstein at the Hebrew University of Jerusalem showed that the information content of a black hole is proportional to the two-dimensional surface area of its event horizon – the point of no return for in-falling light or matter. Later, string theorists managed to show how the original star's information could be encoded in tiny lumps and bumps on the event horizon, which would then imprint it on the Hawking radiation departing the black hole.

This solved the paradox, but theoretical physicists Leonard Susskind and Gerard 't Hooft decided to take the idea a step further: if a three-dimensional star could be encoded on a black hole's 2D event horizon, maybe the same could be true of the whole universe. The universe does, after all, have a horizon 42 billion light years away, beyond which point light would not have had time to reach us since the big bang. Susskind and 't Hooft suggested that this 2D 'surface' may encode the entire 3D universe that we experience – much like the 3D hologram that is projected from your credit card.

It sounds crazy, but we have already seen a sign that it may be true. Theoretical physicists have long suspected that space–time is pixelated, or grainy. Since a 2D surface cannot store sufficient information to render a 3D object perfectly, these pixels would be bigger in a hologram. 'Being in the [holographic] universe is like being in a 3D movie,' says Craig Hogan of Fermilab in Batavia, Illinois. 'On a large scale, it looks smooth and three-dimensional, but if you get close to the screen, you can tell that it is flat and pixelated.'

## Quantum fluctuation

Hogan recently looked at readings from an exquisitely sensitive motion-detector in Hanover, Germany, which was built to detect gravitational waves – ripples in the fabric of space–time. The GEO600 experiment has yet to find one, but in 2008 an unexpected jitter left the team scratching their heads, until Hogan suggested that it might arise from 'quantum fluctuations' due to the graininess of space–time. By rights, these should be far too small to detect, so the fact that they are big enough to show up on GEO600's readings is tentative supporting evidence that the universe really is a hologram, he says.

Bekenstein is cautious: 'The holographic idea is only a hypothesis, supported by some special cases.' A dedicated instrument built at Fermilab in 2014, the Holometer, hoped to find more evidence. It didn't, but that doesn't rule the idea out.

Solid evidence for the holographic universe would challenge every assumption we have about the world we live in. It would show that everything is a projection of something occurring on a flat surface billions of light years away from where we perceive ourselves to be. As yet we have no idea what that 'something' might be, or how it could manifest itself as a world in which we can do the school run or catch a movie at the cinema. Maybe it would make no difference to the way we live our lives, but somehow I doubt it.

## You might be older – or younger – than your years

*Biological age can diverge from the number of years we celebrate on our birthdays, says **Helen Thomson** – and it sheds light on the time we have left.*

Age is a peculiar concept. We tend to think of it as the number of birthdays we have celebrated – our chronological age. But this is just one indicator of the passage of time. We also have a biological age, a measure of how quickly the cells in our body are deteriorating compared with the general population. And these two figures don't always match up.

Just take a look around: we all know people who look young for their age, or folks who seem prematurely wizened. Even in an individual, different parts of the body can age at different speeds. By examining how chronological age lines up with biological age across the population, researchers are starting to pin down how these two measures should sync up – and what it means for how long we have left when they don't.

Studies have shown that our biological age is often a more reliable indicator of future health than our actual age. It could help us identify or even prevent disease by tracking the pace at which we're getting older. It may even allow us to slow – or reverse – the ageing process.

I became interested in my biological age after discovering in my 20s that my ovaries were ageing prematurely. Yet now, at 33, I am still often asked for identification when buying alcohol, suggesting my face is holding up pretty well. It made me wonder about other aspects of my biological age, and whether knowing more might help me to live a longer, healthier life. So, I set out to answer the question: How old am I really?

Ageing is the progressive loss of function accompanied by decreasing fertility and increasing mortality, according to Thomas Kirkwood from the Institute for Ageing at the University of Newcastle. Surprisingly, it's not universal across species. The *Turritopsis dohrnii*, or 'immortal jellyfish', can revert to a larval state and turn back into an adult indefinitely, for instance. We don't have that luxury. According to the UK Office for National Statistics, I can expect to live to 83.

The most widely cited theory of ageing is that telomeres, genetic caps on the ends of chromosomes, grow shorter each time a cell divides – like a wick burning on a candle. Once these are used up, the cell withers and dies. But a new idea gaining ground suggests ageing is instead a byproduct of how energy intensive it is for our bodies to continuously repair faults that occur in our DNA as cells divide. 'It doesn't make evolutionary sense to maintain that process for ever,' says Kirkwood. Indeed, several animal studies have shown that genes that affect lifespan do so by altering cells' repair mechanisms. Little by little, faults build up in cells and tissues and cause us to deteriorate.

This is where biological age comes in – it attempts to identify how far along we are in this process. It's not a simple task, because no one measure of cellular ageing gives a clear picture. As Kirkwood says, 'Attempts to measure biological age have been bedevilled by the difficulty of taking into account the many different biological processes at work.'

Still, a growing number of researchers have taken up the challenge. Before seeking them out, however, I began to wonder whether I could be in for a nasty surprise. When Daniel Belsky and his team at Duke University in North Carolina studied 18 different markers of cellular ageing – including blood pressure and cardiovascular function – in almost 1000 adults, they found that some were ageing far faster or slower than their birth certificates would suggest. One 38-year-old had a biological age of 28; another's was 61.

So if I have an accelerated biological age, does it mean I'm less likely to make it to 83? Studying humans until they die takes a long time, so the causal relationship is tricky to pin down. But an increasing number of studies suggest this is a fair assumption. Belsky's team found that 38-year-olds with an older biological age fared worse on physical and mental tests, for instance. And when James Timmons and colleagues at King's College London examined expression of 150 genes associated with ageing in 2015, they found that biological age was more closely tied to risk of diseases such as Alzheimer's and osteoporosis than chronological age.

Braced for a rocky ride, I started the hunt for my real age by looking in the mirror. In 2015, Jing-Dong Jackie Han and colleagues at the Chinese Academy of Sciences in Shanghai analysed 3D images of more than 300 faces of people between 17 and 77 years old, and created an algorithm to predict age. When they used it on a new group of faces, they found that people born the same year differ by six years in facial age on average, and that these differences increase after 40.

'Some molecular changes in the body can be reflected on the face,' says Han. High levels of low-density cholesterol (the 'bad' kind) are associated with puffier cheeks and pouches under the eyes, for instance. Dark circles under the eyes can result from poor kidney function or blood circulation. The message is that if we look older than we should, it could be a sign of underlying disease.

The algorithm was developed using a population of Han Chinese people and so far has only been tested in four people of white European descent. So, as a white woman, I had my face analysed by a similar algorithm designed by anti-ageing company Youth Laboratories in Russia. The result was a win for me: I apparently have the face of a 25-year-old.

Next it was time to draw some blood. Using 32 different parameters that reflect disease risk, a team at the company Insilico Medicine developed a deep-learning algorithm to predict age. After training it on more than 60,000 blood samples of known chronological age, they used it to accurately predict age from new samples to within 6.2 years. The team found that people whose blood age was higher than their actual number of years were more likely to have health problems. The algorithm is free to use, so after I had my blood taken by Medichecks in London, I plugged in my details at www.aging. ai. Reassuringly, it shaved off a couple of years, estimating my real age to be 31.

Another method for measuring biological age is to look at how complex carbohydrates called glycans are attached to molecules in the body, a process called glycosylation. Gordan Lauc and colleagues at the University of Zagreb recently discovered that glycosylation of an antibody called immuno-globulin G changes as we get older, and that this can be used to predict chronological age. When Lauc's team compared 5117 people's 'glycan age' with known markers for health deterioration, such as insulin, glucose, BMI and cholesterol, they found that those who scored poorly on these markers also had an older glycan age.

'Your glycan age seems to reflect how much inflammation is occurring in the body,' says Lauc. Prolonged inflammation can make cells deteriorate faster, so having an accelerated glycan age could be used as an early warning signal that your health is at risk, he says.

Lauc and Tim Spector, a genetic epidemiologist at King's College London, founded GlycanAge – a company that tests people's glycan levels – and kindly tested mine for free. It turns out my glycan age is just 20, a whopping 13 years younger than I am.

With a new spring in my step, I moved on to what is now considered the most accurate way to measure human ageing: an intrinsic 'epigenetic' clock present in all our cells. Epigenetics refers to the process by which chemical tags called methyl groups are added to or removed from DNA, which in turn influences which genes are switched on or off. Some changes in methylation patterns over time can be used to estimate age.

The father of this technique is Steve Horvath at the University of California, Los Angeles. In 2011, looking at methylation patterns in blood samples, Horvath and colleagues were able to predict chronological age to within five years. He has since analysed data from more than 13,000 samples and identified methylation patterns to estimate a healthy person's age to within 2.9 years. 'The age estimate is so accurate it continues to amaze me,' says Horvath. (Unfortunately, for the purposes of my investigation, at $900 a pop, I decided to give this test a miss.)

Horvath is also interested in discrepancies between our chronological age and epigenetic clock, which diverge most drastically in cancer tissue. Trey Ideker, a medical researcher at the University of California, San Diego, and his colleagues discovered that the epigenetic age of kidney, breast, skin and lung cancer tissue can be almost 40 per cent older than the person it came from.

A recent study by Horvath and his team suggests that breast tissue from healthy women aged 21 appears 17 years older than their blood, which tends to correlate closely with their chronological age. This difference decreases as we get older; for women aged 55 years, breast tissue appears around eight years older than blood. By identifying what the normal differences are, researchers hope to flag outliers. 'Ultimately, we want to be able to collect data from a particular organ, or from a surrogate tissue and say, "Wow, this woman has breast tissue that is 20 years older than it should be, so she needs to be monitored more closely for breast cancer",' says Horvath.

Beyond monitoring and aiding diagnoses for diseases, can any of these measures give us a better idea of how much life we have left? There is an association between our epigenetic clock and our time to death, but it's not very accurate – yet.

In his analyses, Horvath found an association between accelerated epigenetic ageing – an older epigenetic age compared with your real age – and time to death. Around 5 per cent of the people he studied had an accelerated epigenetic age. Their risk of death in the next decade was about 50 per cent higher than those whose epigenetic age lined up with their actual years.

If our epigenetic clock is ticking down to our death, is there anything we can do to intervene? Horvath has started studying the epigenetic age of induced pluripotent stem cells (iPSCs), which are adult cells that can be pushed to revert to an embryonic-like state, from which they are capable of turning into most types of cells in the body.

The epigenetic age of iPSCs is zero. Transforming normal body cells into stem cells would be an 'extreme rejuvenation procedure', Horvath says. You wouldn't want to do it to all of your cells, but perhaps it's a strategy that could be modified to intervene with the ageing process. 'It sounds like science fiction, but conceptually it's possible,' he says. 'All epigenetic marks are reversible, so in theory it's possible to reset the clock.'

## Turn back time

Another promising, if speculative, plan might be to freeze blood stem cells when you are young so that you can use them to reconstitute your immune system when you are old.

Short of miraculous anti-ageing treatments, understanding more about biological age can still improve our health. People told their heart age – measured using parameters such as blood pressure and cholesterol – are better able to lower their risk of cardiovascular problems compared with people given standard information about heart health, for instance. (My heart, I learned, is 28 years old.)

There are not yet any placebo-controlled trials to determine whether certain lifestyle interventions can reduce biological age, and so risk of early death. But Horvath did find that the epigenetic clock is accelerated in the livers of obese people, and ticks more slowly for those who regularly consume fish and vegetables, and only drink in moderation.

Unsurprisingly, exercise also seems to help. In a trial of more than 57,000 middle-aged people, those whose fitness levels resembled a younger person's were less likely to die in the following decade or so. Fitness-associated biological age was a stronger predictor of survival than chronological age.

There is still a long way to go before we can pinpoint the exact ways to reverse ageing. But for now, I'm relieved to know that most of my body is younger than my years would suggest and, in the not too distant future, knowing my biological age could hold the key to preventing disease or even postponing death. I'll happily celebrate turning 34 in the knowledge that my age really is just a number.

## How the story of human origins is being rewritten

*The past 15 years have called into question every assumption about who we are and where we came from. It turns out our evolution is more baffling than we thought.* **Colin Barras** *tries to unpick a very tangled family tree.*

Who do you think you are? A modern human, descended from a long line of *Homo sapiens*? A distant relative of those great adventure-seekers who marched out of the cradle of humanity, in Africa, 60,000 years ago? Do you believe that human brains have been getting steadily bigger for millions of years, culminating in the extraordinary machine between your ears?

Think again, because over the past 15 years, almost every part of our story, every assumption about who our ancestors were and where we came from, has been called into question. The new insights have some unsettling implications for how long we have walked the earth, and even who we really are.

Once upon a time, the human story seemed relatively straightforward. It began roughly 5.5 to 6.5 million years ago, somewhere in an East-African forest, with a chimpanzee-like ape. Some of its descendants would eventually evolve into modern chimps and bonobos. Others left the forest for the savannah. They learned to walk on two legs and, in doing so, launched our own hominin lineage.

By about 4 million years ago, the bipedal apes had given rise to a successful but still primitive group called the australopiths, thought to be our direct ancestors. The most famous of them, dubbed Lucy, was discovered in the mid-1970s and given arch-grandmother status. By 2 million years ago, some of her descendants had grown larger brains and longer legs to become the earliest 'true' human species. *Homo erectus* used its long legs to march out of Africa. Other humans continued to evolve larger brains in an apparently inexorable fashion, with new waves of bigger-brained species migrating out of Africa over the next million years or so, eventually giving rise to the Neanderthals of Eurasia.

Ultimately, however, those early migrant lines were all dead ends. The biggest brains of all evolved in those hominins who stayed in Africa, and they were the ones who gave rise to *Homo sapiens*.

Until recently, the consensus was that our great march out of Africa began 60,000 years ago and that by 30,000 years ago, for whatever reason, every other contender was extinguished. Only *H. sapiens* remained – a species with a linear history stretching some 6 million years back into the African jungle.

Or so we thought.

Starting in the early 2000s, a tide of new discoveries began, adding layer upon layer of complexity and confusion. In 2001 and 2002 alone, researchers revealed three newly discovered ancient species, all dating back to a virtually unknown period of human prehistory between 5.8 and 7 million years ago.

Very quickly, *Orrorin tugenensis*, *Ardipithecus ramidus* and *Sahelanthropus tchadensis* pushed a long-held assumption about our evolution to breaking point. Rough genetic calculations had led us to believe our line split from the chimp lineage between 6.5 and 5.5 million years ago. But *Orrorin*, *Ardipithecus* and *Sahelanthropus* looked more like us than modern chimps do, despite predating the presumed split – suggesting our lineage might be at least half a million years older than we thought.

At first, geneticists made grumpy noises claiming the bone studies were wrong, but a decade later, even they began questioning their assumptions. In 2012, revised ideas about how quickly genetic differences accumulate in our DNA forced a reassessment. Its conclusion: the human–chimp split could have occurred between 7 and 13 million years ago.

## Not so chimp-like

Today, there is no longer a clear consensus on how long hominins have walked the earth. Many are sticking with the old assumption, but others are willing to consider the possibility that our lineage is almost twice as old, implying there are plenty of missing chapters to our story still waiting to be uncovered.

The struggles don't end there. The idea that our four-legged ancestors abandoned the forests, perhaps because of a change in climate conditions, and then adapted to walk on two legs is one of the oldest in human evolution textbooks. Known as the savannah hypothesis, it was first proposed by Jean-Baptiste Lamarck in 1809. Exactly 200 years later, an exquisite, exceptionally preserved 4.4-million-year-old skeleton was unveiled to the world, challenging that hypothesis.

'Ardi', a member of *A. ramidus*, is a jewel in the hominin fossil record. She is all the more important because of the number of key assumptions she casts doubt on. Ardi didn't have a chimp's adaptations for swinging below branches or knuckle-walking, suggesting chimps gained these features relatively recently. In other words, the ape that gave rise to chimps and humans may not have been chimp-like after all.

And contrary to Lamarck's hypothesis, her feet, legs and spine clearly belonged to a creature that was reasonably comfortable walking upright. Yet, according to her discoverers, Ardi lived in a wooded environment. This suggests that hominins began walking on two legs before they left the forests, not after – directly contradicting the savannah hypothesis.

Although not everyone is convinced that Ardi was a forest-dweller, other lines of evidence also suggest we have had the upright-walking story back to front all these years. Susannah Thorpe at the University of Birmingham studies orangutans in their natural environment and has found that they stand on two legs to walk along branches, which gives them better access to fruit. In fact, all living species of great ape will occasionally walk on two legs as they move around the forest canopy. It would almost be odd if our own ancestors had not.

Whether before or after standing on two legs, at some stage our ancestors must have come down from the trees. We can depend on that, at least. Entering the twenty-first century, we knew of just one group that fitted the transition stage: the australopiths, a group of ape-like bipedal hominins, known from fossils found largely in east and south Africa and dating to between 4.2 and 1.2 million years ago. They lived in the right place at the right time to have evolved into humans just before 2 million years ago. Lucy would have shown up in the middle of that period, 3.2 million years ago. Since her discovery, she has served as a reassuring foundation stone on which to build the rest of our hominin family tree, a direct ancestor who lived in East Africa's Rift Valley.

Then, in 2001, researchers unveiled a 3.5-million-year-old

skull discovered in Kenya. The skull should have belonged to Lucy's species, *A. afarensis*, the only hominin species thought to be living in East Africa at the time. But its face didn't fit. It was so flat that it could barely be considered an australopith, says Fred Spoor at University College London, who analysed the skull. He and his colleagues, including Meave Leakey at Stony Brook University in New York, gave it a new name: *Kenyanthropus platyops*.

On the face of it, the suggestion that Lucy's species shared East Africa with a completely different type of hominin seemed only of marginal interest. But within a few years, the potential significance of *Kenyanthropus* was beginning to grow. After comparing the skull's features with those of other hominin species, some researchers dared suggest that *K. platyops* was more closely related to us than any australopithecus species. The conclusion pushed Lucy on to a completely different branch of the family tree, robbing her of her arch-grandmother position.

If that wasn't confusing enough, other researchers were making a similar attack from a different direction. The discoverers of *O. tugenensis*, the 6-million-year-old hominin found in 2001, also concluded that its anatomy was more human-like than that of australopiths, making it more likely to be our direct ancestor than Lucy or any of her kin.

Most of the research community remains unconvinced by these ideas, says Spoor, and a recent announcement that a human-like jawbone 2.8 million years old had been discovered in Ethiopia once more shored up Lucy's position. 'In many respects it's an ideal transitional fossil between *A. afarensis* and earliest *Homo*,' says Spoor.

Even so, Lucy's status as our direct ancestor has been formally challenged, twice, and Spoor says it's not inconceivable that the strength of these or other challenges will grow.

'We have to work with what we have and be prepared to change our minds if necessary.'

## Tiny brains and alien hobbits

Intriguingly, in 2015, a team announced the discovery of the oldest known stone tools. The 3.3-million-year-old artefacts were found in essentially the same deposits as *Kenyanthropus*. 'By all reasonable logic *Kenyanthropus* would be the tool-maker,' says Spoor. Perhaps that hints at a tool-making connection between *Kenyanthropus* and early humans – although there is circumstantial evidence that some australopiths used stone tools too. In any event, determining which hominins evolved into humans is no longer as clear-cut as it once was.

Other important parts of the human evolution narrative were untouched by these discoveries, in particular, the 'out of Africa' story. This idea assumes that the only hominins to leave Africa were big-brained humans with long legs ideally suited for long-distance travel.

But discoveries further afield have begun to chip away even at this core idea. First came news, in 2002, of a 1.75-million-year-old human skull that would have housed a brain of no more than 600 cubic centimetres, about half the size of modern human brains. Such a fossil wouldn't be an unusual find in East Africa, but this one turned up at Dmanisi in Georgia, in the Caucasus region. Clearly, small-brained hominins had left Africa.

In other respects, the Dmanisi skull and several others found at the site did not threaten the standard narrative. The Dmanisi hominins do seem to be early humans – perhaps unusually small-brained versions of *H. erectus*, conventionally regarded as the first hominin to leave Africa.

A discovery in 2003 would ultimately prove far more problematic. That year, researchers working on the Indonesian island of Flores found yet another bizarre skeleton. It had the small brain and small body of an early African hominin, from around 2 to 3 million years ago. To make matters worse, it seemed to have been alive just a few tens of thousands of years ago in a region thought to be home only to 'true' long-limbed and large-brained humans. The team named the peculiar species *Homo floresiensis*, better known by its nickname: the hobbit.

'I said in 2004 that I would have been less surprised if they had found an alien spacecraft on Flores than *H. floresiensis*,' says Peter Brown at the Australian National University, who led the analysis of the remains. The primitive-looking skeleton was, and still is, 'out of place and out of time'.

There's still no agreement on the hobbit's significance, but one leading idea is that it is evidence of a very early migration out of Africa involving prehuman australopith-like hominins. In fact, the entire out-of-Africa narrative is in flux, with genetic and fossil evidence suggesting that even the once widely held opinion that our species left Africa 60,000 years ago is hopelessly wrong. Some lines of evidence suggest *H. sapiens* may have reached China as early as 100,000 years ago.

The hobbit was just one bizarre hominin, and could reasonably be discounted as a simple anomaly. But within little more than a decade of its discovery, two more weird misfits had come to light, both in South Africa.

*Australopithecus sediba* and *Homo naledi* are quite unlike any hominin discovered before, says Lee Berger at the University of Witwatersrand in South Africa, who led the analysis of both. Their skeletons seem almost cobbled together from different parts of unrelated hominins. Significantly, the mishmash of features in the *A. sediba* skeleton, unveiled in 2010, is very different from those in the *H. naledi* skeleton, unveiled in 2015.

*A. sediba*'s teeth, jaws and hands were human-like while its feet were ape-like. *H. naledi*, meanwhile, combined australopith-like hips with the skull of an early 'true' human and feet that were almost indistinguishable from our own.

No other ancient species seems quite as strange – but, as Berger points out, very few other ancient hominins are preserved in so much detail. Perhaps that's just an interesting coincidence. Or perhaps, he says, it's a sign that we have oversimplified our understanding of hominin evolution.

We tend to assume that ape-like species gradually morphed into human-like ones over millions of years. In reality, Berger thinks, there may have been a variety of evolutionary branches, each developing unique suites of advanced human-like features and retaining a distinct array of primitive ape-like ones. 'We were trying to tell the story too early, on too little evidence,' says Berger. 'It made great sense right up until the moment it didn't.'

In 2017, Berger announced the age of the *H. naledi* remains. They are just 236,000 to 335,000 years old. Weeks later, news broke that 300,000-year-old fossils from Morocco might belong to early members of *H. sapiens*. If correct, the fossil extends our species' history by a whopping 100,000 years.

*H. naledi*'s relatively young age is also a striking example of how complex and confusing the human evolutionary tree might really be. Human brains didn't grow and grow for millennia, with smaller-brained species falling to the wayside of the gradual evolutionary road. Instead, our species occupied an African landscape that was also home to humans with brains half the size of theirs.

We can only speculate on how (or whether) the small-brained

*H. naledi* interacted with the earliest *H. sapiens*. Tantalising but controversial evidence from Berger's team suggests that *H. naledi* intentionally disposed of its dead – perhaps a sign that even 'primitive' hominins could behave in an apparently sophisticated way.

Another independent line of evidence suggests that different behaviour was not necessarily a barrier to inter-species interactions.

In the late 1990s, geneticists began to show an interest in archaeological remains. Advances in technology allowed them to sequence a small chunk of mitochondrial DNA (mtDNA) from an ancient Neanderthal bone. The sequence was clearly distinct from *H. sapiens*, suggesting that Neanderthals had gone extinct without interbreeding ('admixing') much with our species.

But mtDNA is unusual. Unlike the nuclear DNA responsible for the bulk of human genetics, it passes intact from a mother to her children and doesn't mix with the father's genes. 'Mitochondrial DNA is the worst DNA you can choose to look at admixture,' says Johannes Krause at the University of Tübingen in Germany.

By 2010, a very different picture was emerging. Further advances in technology meant geneticists such as Krause could piece together a full nuclear genome from Neanderthal bones. It showed subtle but distinct evidence that Neanderthals had interbred with our species after all. The behavioural differences between humans and Neanderthals were evidently not enough to preclude the occasional tryst.

Arguably, this wasn't the biggest genetics announcement of the year. In their searches, Krause and his colleagues had examined genetic material extracted from a supposed Neanderthal bone fragment unearthed in a Siberian cave in 2008. To everyone's surprise, the DNA in the bone wasn't

Neanderthal. It came from a related but distinct and entirely new hominin group, now dubbed the Denisovans.

To this day, the Denisovans remain enigmatic. All that we have of them are one finger bone and three teeth from a single cave. We don't know what they looked like, although *H. sapiens* considered them human enough to interbreed with them: a Denisovan nuclear genome sequence published in 2010 showed clear evidence of sex with our species. The DNA work also shows that they once lived all across East Asia. So where are their remains?

## Slap and tickle

Fast-forward to 2017, and the interbreeding story has become more complex than anyone could have imagined in 2000. Krause reels off the list. 'Neanderthals interbred with *H. sapiens*. Neanderthals interbred with Denisovans. Denisovans interbred with *H. sapiens*. Something else that we don't even have a name for interbred with Denisovans – that could be some sort of *H. erectus*-like group . . .'

Although weird bones have done their bit to question our human history, it's the DNA inside them that may have done the most to shake up our evolutionary tree. With evidence of so much ancient interbreeding, it becomes far more complicated to decide where to draw lines between the different groups, or even if any lines are justified.

'How do you even define the human species now?' says Krause. 'It's not an easy discussion.' Most of us alive today carry inside our cells at least some DNA from a species that last saw the light of day tens of thousands of years ago. And we all carry different bits – to the extent that if you could add them all up, Krause says you could reconstitute something like one-third of the Neanderthal genome and 90 per cent of the Denisovan genome. With this knowledge, can we even say that these species are truly extinct? Pushing the idea one step further, if most living humans are a mishmash of *H. sapiens* DNA with a smattering from other species, is there such a thing as a 'true' *H. sapiens*?

Having dug ourselves into this philosophically troubling hole, there's probably only one way to find our way out again: keep digging for fossils and probe them for more DNA.

# 3 Life at the Extremes

The universe is overwhelmingly inhospitable to life. Most of it is made up of freezing emptiness lashed by blistering sprays of cosmic radiation. Only tiny pockets can hope to sustain life, cradled against the desolation of deep space. But even the relatively warm, wet, Earth can be far from comfortable. Buried under mountains or sheets of ice, or blown into freezing limbo in the skies, tiny creatures eke out an existence in the most extraordinary conditions. Learning more about them helps us to understand our own world, and also where living creatures might pop up next. Whether on Earth, or maybe even the moons of Jupiter, life finds a way.

# Two tiny dots that defy the history of the solar system

*When did life on Earth begin?* **Colin Stuart** *discovers how two specks of rock that formed when our planet was young suggest we've got to rethink everything from the origins of life to the story of our solar system.*

Of the 200,000 shards of rock that Mark Harrison has retrieved from Australia since the mid-1980s, only one contained what he was looking for. Two flecks of graphite, each barely the size of a red blood cell. Small, perhaps, but capable of overturning everything we know about life on Earth. Harrison, a geologist at the University of California, Los Angeles, remembers thinking to himself: 'By golly, they're a dead ringer for a biogenic origin.' Biogenic means made by life – but how? These graphite flecks were found in a zircon crystal that had lain trapped deep in the Jack Hills of Western Australia for 4.1 billion years. So they seem to imply our planet was inhabited at least 300 million years earlier than anyone had previously imagined.

What's more, these first living organisms would date from a time before our planet was thought capable of harbouring any life at all. In these early years, Earth was supposedly a molten hellhole racked by volcanism and bombarded by space debris, zinging around a solar system yet to find inner peace. If Harrison's fossils are all they seem, they wouldn't only rewrite the history of life and Earth – but the entire solar system's as well.

When it came to explaining how these things all got started, we thought we had it more or less worked out. Some 4.6 billion years ago, a vast cloud of dust and gas in some corner of an unremarkable galaxy began to collapse into a dense ball of matter. As more and more surrounding material was pulled towards it, the temperature and pressure at its core increased, to the point where nuclear fusion kicked in. This released vast quantities of energy and marked the moment our sun became a star.

As the newborn star slowly began to spin, smaller bodies started to coalesce in orbit around it. Close in, vast quantities of water ice were boiled away, leaving only metallic compounds behind to form the smaller rocky planets. Further out, cooler temperatures allowed giant worlds of ice and gas to form. All in a single plane along smooth, near-circular tracks.

It was a nice story, but as further details emerged, it became apparent that this picture was incomplete. For one thing, it struggled to explain the quantity and distribution of the so-called Trojan asteroids, thousands of tiny bodies that chase after Jupiter in its orbit. The Kuiper belt, the icy band beyond Neptune that Pluto belongs to, was equally difficult to justify: many of its bodies orbit at far greater angles to the planetary plane than the conventional picture would allow. Perhaps most perplexing of all, however, was the evidence our cosmic neighbourhood had once been under heavy bombardment. Rocks returned to Earth by the *Apollo* astronauts suggested the widespread cratering on our own moon was the result of a protracted assault which took place 3.9 billion years ago – a ruction the conventional model found hard to explain.

The solution, named after the city in France where it was devised in 2005, was the Nice model. In this refinement of the traditional story, our solar system's four giant planets started out much closer together than they are today. This configuration was unstable, leading to hundreds of millions of years of gravitational tussling, during which the giant planets migrated into their current positions, disturbing the millions of tiny bodies littering the ancient solar system. Many fell under Jupiter's gravitational influence, becoming its Trojan followers, while others settled in the solar system's outer regions as highly angled denizens of the Kuiper belt.

Meanwhile, asteroids in the band between Mars and Jupiter were dislodged from orbit, many going on to collide with the innermost planets. This period of intense activity, known as the Late Heavy Bombardment (LHB), would have left deep craters on the moon and given our fledgling planet a serious knock during the turbulent early stages of its development.

The small number of surviving solid rocks from this period have led us to picture early Earth as a fiery world covered in volcanoes bursting through a molten crust. The LHB's few hundred million years of constant collisions contributed to a nightmarish landscape so extreme that the geological period is known as the Hadean, after the Greek god of the underworld. The existence of life in such a hellscape was considered preposterous. Instead, the first traces of biogenic carbon, dated at 3.8 billion years old, neatly coincide with the time Earth was finally at peace and the bombardment from outer space had slowed.

Hence the excitement if Harrison's fleck of graphite really is what it appears to be: evidence not only of our planet's oldest known life form, but one that emerged at an impossible time. His smoking gun was the ratio of isotopes carbon-13 and carbon-12 within the sample. 'If you were looking at this carbon ratio today, you would say it was biogenic,' he says.

Astonishing as it is, Chris Ballentine from the University of Oxford cautions against getting carried away. 'It is one inclusion in one zircon,' he says. 'But this sets the bar for people to find more and really show there was life around back then.'

Life or no life, it's just the latest piece of evidence from the Jack Hills suggesting Earth's hellish youth was more short-lived than astronomers thought possible. As far back as 1999, geologists uncovered other zircons in this astonishing terrain that indicated part of Earth's surface had cooled and solidified 4.4 billion years ago. What's more, measurements of how much oxygen the rocks contained suggested that Earth had been mild enough to support liquid water.

Further evidence that not all was right in the established picture of Earth and the solar system came in 2013, when Judith Coggon, then at the University of Bonn, was analysing another contender for the planet's oldest rock – on the other side of the world in Greenland. There she found evidence that Earth contained significant quantities of gold and platinum as far back as 4.1 billion years ago – even though these metals were thought to have been delivered only later by the Late Heavy Bombardment.

Yet more contention came in 2015, when Nathan Kaib from the University of Oklahoma, along with John Chambers from the Carnegie Institution in Washington DC, published the results of their latest simulations of solar system formation. What they found seemed to sound the death knell for the Nice model. In 85 per cent of cases, the inner solar system ended up with fewer than the four rocky worlds it has today. 'More often than not you lose Mercury,' says Kaib. Only 1 per cent of the time could they create a solar system that looked like the one we recognise. It would not be the first time the Nice model has been modified to take account of problems, but this was a problem of a different magnitude. 'It seems very unlikely that you can get the outer solar system architecture and protect the inner planets,' he says.

Kaib has a surprisingly simple solution. The giant planets still migrated, producing the Jovian Trojans and the Kuiper belt, but they did so much earlier – while the innermost planets were still forming. By turning up to the party fashionably late, Earth dodged a bullet. The early migration of the giant planets would have scattered most of the larger impactors by the time Earth's formation was complete. That works well, says Zoë Leinhardt, from the University of Bristol. 'The latter part of Earth's formation would have been calmer, as opposed to having formed and then being smacked upside the head.'

It's an appealing theory, explaining not only why the solar system looks the way it does, but how Earth became friendly to life so early. But one final mystery remains. If the giant planet migration happened before Earth and the moon had formed, then something else must have been responsible for the craters on the lunar surface. But what?

David Minton from Purdue University thinks the answer lies closer to home. 'In the Nice model, most of the LHB impactors come from the asteroid belt,' he says. 'But the distribution of crater sizes on the moon and the distribution of asteroids don't match.' Matija Cuk of the SETI Institute agrees. 'If the LHB really was just asteroids being thrown at the moon en masse, there should be a lot more big lunar basins, and there aren't,' he says. Minton believes he might have found an alternative source for the LHB: Mars.

He's still working on the finer details, but he presented the concept to the American Astronomical Society's Division on Dynamical Astronomy at their meeting in May 2015. One fact working in its favour is that the Red Planet's northern hemisphere is low-lying and considerably flatter than the highlands in the south. 'Many have suggested that's because the northern area is a giant basin formed by a 2000-kilometre impactor,' says Minton. Debris thrown up by the formation of this so-called Borealis basin could have bombarded the moon, and Earth, 3.9 billion years ago.

Cuk has an even more radical explanation. 'To me it is not clear at all that there was a spike in lunar bombardment 3.9 billion years ago,' he says. The *Apollo* samples that led to the assumption were returned from several different sites on the moon, with many showing evidence of impacts clustered around that time. But Cuk believes the *Apollo* samples all came from the impact or impacts that formed the Imbrium basin – one of the large, dark patches that makes up the 'Man in the Moon'. Rocky shrapnel from this event could have contaminated disparate parts of the lunar surface, meaning that what at first looked like a host of simultaneous impacts might have only been a handful. 'The idea of the carpet-bombing of the moon 3.9 billion years ago has gone away,' he says. If you could prove the impacts that caused the cratering on the moon were less of a spike and more of a steady drip, then the Nice model could be saved after all. Just as crucially, it would have profound implications for conditions on our infant planet. 'If the impacts were more smeared out, early Earth wouldn't have been total hell,' says Cuk.

Either way, with relative calmness kicking in sooner in Earth's history, life could have emerged more quickly to leave its mark in the Jack Hills zircon. 'Pushing giant planet migration back earlier would be consistent with what we found,' says Harrison. Future work will look at cementing this idea.

If Harrison's hunch is right, then the life forms we had

previously thought of as our earliest ancestors, dating from 3.8 billion years ago, weren't the beginning of the evolutionary tree at all. Instead, life on Earth began hundreds of millions of years earlier, almost as soon as the planet was ready for it. Such a scenario would raise hopes for the speed and ease with which biology can take hold, and of its aptitude for sticking around in an unfriendly cosmos. According to Harrison, 'it makes the notion of life elsewhere in the universe that much more likely.' Our revised history could point to a more interesting future.

## Giant viruses could rewrite the story of life on Earth

*They're huge and they lurk everywhere. But don't fear, these giants seem to be gentle, says* **Garry Hamilton** *– and might even be ancestral forms of life that taught our own cells a few neat tricks.*

They found the mystery microbe in a water tower in Bradford in 1992. This city in northern England is not, perhaps, the first place you'd expect to find exotic life forms. But whatever it was looked bizarre under a microscope. It was a hairy, 20-sided polyhedron, which hinted that it was a virus. But it seemed far too big for that, And closer inspection revealed a complexity totally out of line with what biologists thought possible for these infectious agents that inhabit the shadowy borders of life.

Yet virus it was. And since 'mimivirus' was eventually confirmed as such in 2003, discoveries of surprisingly enormous viruses have kept coming. Not only do giant viruses seem to be all over the place, but their world is more vast and diverse than we ever imagined. Their genes are hinting that we need to come up with a new classification of life and are revealing that viruses might just have had a hand in making us who we are. 'If we want to understand evolution and the origins of life, viruses have to be taken into account,' says Patrick Forterre, a molecular biologist at the Pasteur Institute in Paris.

Biologists have been loath to allow viruses the prestige of being labelled as alive because they can do little without a host. They are parasites, injecting their genetic instructions into a cell and hijacking its biochemical machinery to produce the parts for spawn. Many of them do little more than replicate, which means they don't need many genes. Human immuno-deficiency virus (HIV), for instance, has just nine. It turns out that mimivirus, in contrast, has 1018.

The other piece of received wisdom about viruses is that they are small. This assumption dates back to 1892, when they were discovered by Dmitri Ivanovsky, a Russian botanist puzzling over an unknown disease ravaging tobacco crops. He filtered the sap from diseased plants through porcelain and found that it remained infectious. Since the pores in the filters were smaller than any bacterium, the conclusion was that the sickness must have been caused by something much tinier. They were later named viruses.

It was partly because of these preconceptions that giant viruses escaped detection for so long. Mimivirus was collected from that Bradford water tower when researchers were looking for the source of a pneumonia outbreak at a nearby hospital. But it was dismissed as just another unclassifiable bacterium, stuck in a freezer and forgotten.

In 1998, a French scientist named Bernard La Scola took another look. He became intrigued because the microbe didn't have any ribosomes, the factories that make proteins and are a hallmark of all cellular life. But the clinching evidence came when a group of scientists showed that the entity didn't divide its cells to reproduce, as all bacteria do. This was definitely a virus.

Over the next decade, Abraham Minsky of the Weizmann Institute of Science in Rehovot, Israel, started delving further into what makes mimivirus tick. Among his first discoveries was a peculiar five-armed star shape that looked almost like it had been tattooed on the virus. 'We saw this amazing fivefold structure,' Minsky says, 'and we had no idea whatsoever what it was.'

It turned out to be the seam of a portal consisting of five triangular panels that swing outwards during infection, allowing the contents of the viral particle to be released into the host. Nobody had seen anything like it before. Minsky named it the 'stargate'.

Minsky also looked at the virus factory that mimivirus creates. Several conventional viruses were already known to set up such an assemblage, which acts like a workshop inside the host cell that churns out virus progeny. None, however, is quite like this one. The mimivirus staffs its factory with the cell's own ribosomes, which make proteins, and mitochondria, which provide power. It's also huge compared with other known factories, big enough to hold hundreds of new viruses. 'We still do not know what the mechanism is by which the ribosomes, mitochondria and so on are recruited,' says Minsky. 'But clearly it's a very efficient, very directed process.'

Finding out how giant viruses work now feels like a more urgent quest, especially since we've discovered that they are just about everywhere. One of those who has helped confirm this is Jean-Michel Claverie, a virologist at Aix-Marseille University, who was part of the team that first identified mimivirus in 2003. 'I suspect there are more very large viruses that have escaped detection,' he said at the time. 'When you think about it, there really is no limit for how big a virus can be.' Since then, he and his wife Chantal Abergel have been on a mission to find more specimens.

The first discovery, however, went to another researcher at the same university, Didier Raoult. He began searching in the most obvious place: more water towers. And sure enough, he struck gold in Paris, discovering a specimen he named mamavirus. But the big news was what they found inside it: the Sputnik virophag. It was the first ever sighting of a virus that infects another virus. (Sputnik translates from Russian as 'fellow traveller'.) The discovery provided fuel for the argument that viruses are somehow alive, since evidently mamavirus can get 'sick'.

Then in 2010, Raoult published the results of a wider search showing that new strains – 19 in total – cropped up in other water samples taken from rivers, lakes, fountains and taps. Things mushroomed from there. The following year, Claverie found an even bigger virus – he named it megavirus – in the ocean off Chile. And in mud samples taken from a river in Chile and a pond in Australia, he and his team unearthed two examples of what is now known as pandoravirus. One had about 1500 genes – the other had more than 2550.

Then came perhaps the most impressive specimen. In ice core samples that froze 30,000 years ago, Claverie found the spectacular pithovirus. At roughly 1.5 micrometres long, this beast is as big as some common bacteria. It also has weird features, including a hole in its membrane that is capped by a 'cork'.

'Now we realise giant viruses are basically everywhere,' says Claverie. 'I'm sure if we looked with the right methods, we would find them in your garden.' They also turn up inside us. The question now bothering biologists is: where did these things come from, and where do they fit into the established classification system for life?

At its coarsest level, this system consists of eukaryotes, bacteria and archaea. Eukaryotes are cells like those that make up animals and plants, with their DNA neatly encapsulated in a nucleus. Bacteria cells are simpler and don't have a nucleus. Archaea are similar to bacteria, but are built on different chemistry. These three fundamental divisions were thought to encompass all living organisms.

The strange thing about giant viruses is that when you take a peek at their genes, they don't seem to fit in anywhere. For any given giant virus, between 50 and 90 per cent of their genes are not known anywhere else. Even the different families of giant virus do not share many. How can this be?

Claverie has a radical suggestion: that giant viruses are the remnants of long extinct domains of life that were completely different from the cells that exist today. It is domains, plural, he says, because the giant viruses are all so different.

## Beyond bounds

'With the mimivirus we argued that we need to invent a fourth domain of life,' he says. 'Now we believe it is no longer just a fourth domain, but a fifth, a sixth and a seventh.' His idea is not without support. In 2012, a team led by Gustavo Caetano-Anollés of the University of Illinois at Urbana-Champaign created an evolutionary tree based on grouping viruses and cells that have similar protein structures. This makes giant viruses seem to be more ancient than anything else, supporting Claverie's theory that they come from extinct lineages.

On the other hand, many argue that the supposedly unique

genes aren't what they seem. They think that viruses evolve much faster than cells, so if you see a gene you don't recognise, it's more likely to be a familiar one mutated beyond recognition than evidence of an unknown domain of life. 'There is absolutely no indication that the giant viruses originated from an extinct or unknown domain of cellular life,' says Eugene Koonin, an evolutionary geneticist at the National Center for Biotechnology Information in Bethesda, Maryland.

Yet Claverie remains resolute. The idea that viruses evolve more quickly than cells doesn't necessarily apply to viruses that use DNA rather than RNA, as the giant viruses do, he claims. And his own studies show that giant virus genes don't evolve faster than host genes. He also argues that all genomes should get smaller, not larger, once they adopt a parasitic lifestyle, because they make use of the host's resources. That means giant viruses must be older than the small viruses we're familiar with, he reckons.

Whatever the truth, viruses clearly have their own way of doing things. 'They are certainly playing around with evolution more than anything else on the planet,' says Curtis Suttle, a marine virologist at the University of British Columbia in Vancouver. One example is the discovery that mimivirus has its own way of making collagen, a protein that crops up in everything from skin to tendons and makes up roughly a quarter of the weight of proteins in most mammals.

But if they are engines for genetic novelty, viruses are also generous with their creations, spreading genes and influencing the evolution of their hosts. So far, we have mostly found giant viruses that infect amoebas. But in 2014, Jonathan Filée, an evolutionary geneticist at Paris-Sud University in France, showed that 23 core giant virus genes can be found in a selection of cellular organisms, including a moss and a gelatinous freshwater animal known as a hydra. It was an indirect pointer that giant viruses infect them too.

'For a long time, we thought viruses were stealing genes from the host,' says Claverie. 'Now it's becoming clearer that viruses often transmit genes to their hosts.' This process might have had serious impacts on the evolution of cells that went on to become life like us. Take the nucleus inside all our cells, for instance. For decades, scientists have kicked around the idea that this was a virus that never left. The idea is largely based on speculation, yet giant viruses might finally provide some backing for it.

In 2013, a team in the US showed that the mimivirus factory is built from the same stuff as the nucleus of the infected amoeba. They also happen to be the same size. It's far from solid proof, but some see this as a hint of an evolutionary link. 'Large DNA viruses probably played an important role in the emergence of eukaryotes by bringing many new genes,' says Forterre. It is possible, he thinks, that the ancestors of modern eukaryotic cells learned how to manipulate membranes and make a nucleus from viruses.

This tangled fallout from the discovery of giant viruses is also changing the way that Forterre, Claverie and others think of life. They say that a virus should not be defined by its inert particle phase, but by the form it takes when united with the genome of its host. In this state, they argue, a virus resembles a parasitic bacterium and is alive. Of course, this living organism isn't anything like the ones we're used to, which is why they also say we need to broaden our thinking. Defining life only as autonomous, ribosome-bearing cells 'is totally too rigid', says Claverie. And Raoult has proposed a division of life into ribosome-bearing 'ribocells' and virus-driven 'virocells'.

Semantics aside, it's clearer than ever that the boundary

between life and viruses is a blurry one. 'It's no longer possible to say that viruses are not living and cells are living,' says Forterre. What comes next is anyone's guess. The search for giant viruses has so far centred largely on amoebas, partly because they are a known host that is straightforward to grow and study in labs. That means there are multitudes of potential hosts still to sift through. For Claverie, that's a daunting and exciting prospect: 'We don't know what a virus is any more – or what to expect next.'

## The lost world hidden under Antarctica's ice

*A buried land of lakes, rivers, volcanoes and even life is changing our view of Earth's seventh continent.* **Anil Ananthaswamy** *plunges beneath the last great wilderness on Earth.*

It took three weeks of crossing frozen terrain to reach the lake, and five days to punch a hole through its icy lid. When they finally broke through to the water below, the excitement was palpable. Hands grabbed the gooey mud pulled up through the hole. For this was no ordinary ice-fishing exped-ition: Slawek Tulaczyk and his team had drilled through 800 metres of ice into Antarctica's Lake Whillans.

The team's efforts – battling through 14-hour shifts in some of the harshest conditions on Earth – are part of a massive endeavour to uncover the continent's hidden secrets. Over a century ago, explorers trudged across its white blanket in pursuit of world records, aware only of the snow, ice and treacherous weather. But in the last few decades a different kind of explorer has started to peer beneath the ice, to discover what Jill Mikucki at the University of Tennessee in Knoxville describes as a sub-ice water-world. Their labour has revealed a pulsating continent with lakes, rivers, volcanoes, even life: hardly the frozen wasteland of popular imagination.

In a way, the adventure began in 1957, when a ship carrying members of the third Soviet Antarctic expedition arrived at the East Antarctic ice sheet. Their aim was to establish a base near the Pole of Inaccessibility, the furthest point on the continent from the Southern Ocean. Thirty-two men struck out from the coast with the equipment that was left after a storm broke up the ice around their ship while they were unloading – sinking sledges and a tractor but no men.

At regular intervals along the way, they set off small explosives and recorded the echo of seismic waves as they travelled through the ice and bounced off whatever lay beneath. Near the centre of East Antarctica, the explorers found a region of anomalously thin ice. They had stumbled across a massive mountain range beneath, its peaks reaching up towards their feet. Almost 3000 metres high, the Gamburtsev mountain range has an Alpine topography, replete with rugged peaks and hanging valleys, yet it is completely hidden from view. The ice above is anywhere from a few hundred metres to 3.2 kilometres thick.

Then, during the late 1960s and early 1970s, planes equipped with ice-penetrating radar revealed bodies of water locked between the ice and the bedrock: lakes hundreds, sometimes thousands of metres beneath, and still liquid thanks to the immense pressure of the ice above and geothermal heat from below. These were the days before GPS and its Russian equivalent GLONASS, and pilots had to keep the Soviet Vostok research station in their sights or risk getting lost over the vast, white, featureless expanse. So it was pure luck that beneath the station sat the continent's biggest lake: Lake Vostok, seventh largest in the world by volume, fourth deepest and 3.7 kilometres under the ice.

Vostok – like Antarctica's other large lakes – sits in a depression in the bedrock and is 'inactive': it fills and drains very slowly. But in recent years, teams studying other subglacial lakes have discovered a dynamic system of streams and even rivers that interconnects some of them.

Duncan Wingham of University College London and his team were the first to spot the massive movement of water beneath the ice. In 2006, they showed how parts of the East Antarctic ice shelf rose and fell, as if the ice were breathing. When the ice sank in one location, a similarly abrupt rise was seen several hundred kilometres away. They concluded that water was flowing from one set of buried lakes to another. 'That was quite a revelation,' says Hugh Corr of the British Antarctic Survey (BAS).

At the last count, polar researchers have identified about 400 subglacial lakes in Antarctica. The discovery has turned the image of a massive ice sheet grinding against the bedrock on its head. Rather, there is an entire hydrological system between the ice and rock. 'Just how much water there is underneath has been a surprise,' says Corr.

There's fire down there, too. During the 2004–05 Antarctic summer, a joint US–UK team carried out an airborne survey to take radar, magnetometer and gravity measurements near Pine Island glacier in West Antarctica. They found something hundreds of metres beneath the ice surface that strongly reflected their radar signal. Corr and David Vaughan, also at BAS, analysed the findings and concluded that the reflections were coming off a layer of ash and rocks, the remnants of a massive volcanic eruption.

The volcano, known as Mount Casertz, erupted about 2000 years ago, punching its way through the ice to spread debris over 26,000 square kilometres. The explosion would have been on the scale of the 1980 blow-up of Mount St Helens in the US. Even today, the ice over Mount Casertz is depressed, suggesting heightened geothermal activity underneath. And in 2010 and 2011, seismometers picked up the rumblings of another active volcano in West Antarctica.

It was against this backdrop of fire and ice that Tulaczyk, of the University of California in Santa Cruz, set out for Lake Whillans in West Antarctica. In December 2012, an advance party of tractors – each dragging a chain of shipping containers mounted on sledges – had to cover the 800 kilometres separating the McMurdo research station from the lake. The containers held nearly 500 tonnes of equipment. Designed to be stacked on ships, not pulled across undulating, wind-hardened ice and snow, they didn't travel well. The crew had to keep welding fractured metal, sometimes cutting parts from one container to patch up another.

By the time the convoy reached the remote site and Tulaczyk and 60 more scientists had flown in from McMurdo, it was mid-January. They had until the end of the month before temperatures would begin to fall. In that time they had to drill through 800 metres of ice to Lake Whillans, drop their instruments through the hole one by one, take measurements, gather samples, then pack everything up and head home.

Tulaczyk remembers the day they finally broke through and dredged up mud from the lake bed. 'People were grabbing the stuff as a souvenir,' he says. 'After spending weeks and weeks surrounded by just snow, to have this very tactile, physical evidence that there is something else [besides] ice underneath our feet . . . it was so amazing for them.'

One of the instruments dropped down the hole was a sensor to take the lake bed's temperature. Until then, there had been only indirect hints that Antarctica's underbelly was warm. 'The measurement came out to be extremely hot – Yellowstone hot,' says Tulaczyk. Similar measurements have been taken at some 35,000 sites around the planet. Only 100 or so are hotter than the Whillans lake bed.

We don't yet know why West Antarctica is so hot and volcanic. It could be that the crust is thinning there. As an analogy, Tulaczyk points to the area between the Sierra Nevada mountains in the western US and Salt Lake City in Utah: over time, the two have moved apart, stretching Earth's lithosphere and producing geothermal and volcanic activity.

Or, paradoxically, the heat could be generated by the ice. The West Antarctic ice sheet has grown and shrunk many times over the past few million years. When the ice is thicker, it depresses the crust, which rebounds when the ice melts. As the crust bobs up and down, so does the viscous mantle 100 kilometres lower down. This constant massaging of the mantle can release heat.

## Life ahoy!

Whatever its source, geothermal heat means liquid water – and that could, in turn, mean life. In the McMurdo Dry Valleys, bright-red brine flows from cracks in the Taylor glacier on to the frozen surface of Lake Bonney. The striking colour of Blood Falls, as the site is known, is the result of iron particles that oxidise in the sunlight. Inside the brine, Mikucki and colleagues have found evidence of 'chemoauto- trophic' bacteria that live in complete darkness – presumably beneath the Taylor glacier – by chemically leaching energy from the bedrock and producing iron as a by-product. They don't rely on energy from the sun, not even in the indirect way that, say, fish at the bottom of the ocean do when they eat dead material that falls from the surface.

The team now want to find where the bacteria come from. Blood Falls is a salty waterfall, and salty water conducts electricity better than fresh water. So in 2015, Mikucki and her team went out to the area with instruments that can remotely measure electrical conductivity beneath the ice. They found large pockets of high conductivity beneath the Taylor glacier. The researchers believe the sediments there are saturated with brine and could host the microbial ecosystems that flow out at Blood Falls.

Although soggy sediments that can support microbial life are definitely prized findings, the treasured goal for Antarctic researchers is to find life in liquid subglacial lakes. Over the past decade Russian scientists have been drilling down to Lake Vostok. The first time they broke through to it, in February 2012, the samples they brought back up to the surface were badly contaminated with the drilling fluid they used to keep the borehole open. After colleagues in Grenoble, France, cleaned them to remove any trace of contaminants, the Russians found no conclusive signs of microbial life.

When they broke through to the lake a second time, in January 2015, special care was taken to prevent contamination. But with their current drilling equipment, it's hard to avoid contamination entirely. Any search for life under these conditions is bound to be inconclusive.

'We need clean water samples to make conclusions. We dream about it,' says Irina Lekhina of the Arctic and Antarctic Research Institute in St Petersburg. She would also like to have water from the main body of the lake. At the minute, they are only sampling the very top layer, which rises into the borehole during the final phase of drilling. 'We believe that life, if it exists, will be deeper,' says Lekhina.

Tulaczyk and his colleagues did get a sample of pristine water from Lake Whillans. Admittedly, their task was easier, with far less ice to get through and warmer temperatures to contend with. They didn't need to keep the borehole open with drilling fluid. Instead, they used a hot-water drill – collecting and boiling snow at the surface, irradiating it with UV light to kill everything inside, and then pumping it down through the ice. The nozzle and other equipment were also irradiated and washed with hydrogen peroxide. When the team broke through to Whillans on 27 January 2013, it was the first clean drilling into a subglacial lake, says Tulaczyk.

In the water samples they found plenty of evidence for a thriving microbial ecosystem. DNA sequences suggest that the microbes in Lake Whillans are chemoautotrophs, like the ones at Blood Falls. 'They get their energy from rocks and mud and stuff underneath the ice,' says team member Ross Powell of Northern Illinois University in DeKalb.

On 8 January 2015, the team drilled at another location, downstream of the lake. This time the borehole went through the Ross ice shelf, just above the glacier's grounding zone, where the ice lifts off the bedrock and begins floating on ocean water. The researchers lowered a camera down into what is essentially an estuary hidden beneath the ice. They found a lot more than microbes, netting shrimp-like amphipods and spying eel-like fish on their screens. Powell wants to go back and trap them. 'It's the ultimate form of ice-fishing,' he says.

The big question is what supplies the energy for all this life. The ocean beneath the ice here is dark and 800 kilometres from open water, so the sun is unlikely to be the energy source. Could the entire food chain – all the way up to amphipods and fish – rely on chemoautotrophic microbes?

If that's the case, it will be only the second chemosynthetic ecosystem known on the planet, says Tulaczyk, the first being deep-sea hydrothermal vents. Despite being devoid of sunlight, vent ecosystems sustain large animals like tube worms – and it all rests on bacteria that get their energy from chemicals spewing out of the vents. Geothermal and volcanic activity beneath the West Antarctic ice sheet could supply energy for large organisms too, not just microbes. 'We haven't found that in Antarctica yet, but it's a possibility that somewhere beneath the ice you are going to find something similar,' says Tulaczyk.

The researchers also want to explore a channel in the bedrock that may carry water from Lake Whillans to the estuary, as it could be bringing nutrients to the area. Powell's team plans to lower a 7-metre-long remotely operated underwater vehicle, bristling with instruments, down a borehole to find out more. 'It's a cigar-shaped cylinder when it goes through the hole, and once it gets into the water it goes through a transformer sort of thing and opens up, and swims about in an open configuration,' says Powell.

It's not just about finding new life forms here on Earth. Exploring subglacial Antarctica may boost efforts to look for life elsewhere in the solar system, such as on the frozen moons Europa and Enceladus. Both harbour liquid water under their vast icy surfaces. 'We don't expect to find a photosynthetically driven ecosystem there,' says Tulaczyk. 'So having a continent that has a chemosynthetically driven system beneath the ice is a nice analogue to pretty much everywhere else that we hope to go and look for life that hasn't originated on Earth.'

We have come a long way from the early days of Antarctic exploration. Less than 150 years ago, geologists thought Antarctica's ice was anchored to the peaks of a volcanic archipelago. No one suspected it hid a continent, let alone life. Our image of Earth's seventh continent has changed forever. 'It's come alive,' says Tulaczyk.

## The unexplored ecosystem above your head

*We have nature reserves on land and at sea, writes* **Lesley Evans Ogden**, *but the sky has never been considered a habitat, let alone one worth preserving, until now.*

The Federal Bureau of Investigation has a spectacular view of the city skyline from its Chicago office tower. But when special agent Julia Meredith arrived at work one Monday morning, her eyes were focused firmly on the ground. That's where the bodies were – more than 10 of them.

Some of the dead were Blackburnian warblers, birds with bright yellow and orange plumage that are rarely seen in the city. They had been on their way to their wintering grounds in South America when they had collided with the building's glass facade. 'They had come all this way, and here they were, dead,' says Meredith.

It's not an isolated incident. In May 2017, 395 migrating birds were killed in a single building strike in Galveston, Texas. The world over, wherever humans are extending their buildings, machines and light into the sky, the lives of aerial creatures are at increasing risk. We don't have very accurate figures, but in the US casualties are thought to run into the hundreds of millions every year. Yet while efforts to protect areas on land and in water have accelerated since the 1970s, the sky has been almost entirely ignored.

That could be about to change if a new wave of conservationists have their way. They want to reclaim the air for its inhabitants, creating protected areas that extend into the sky and designing buildings to avoid death. If this noble aim is to succeed, however, we must first address a more fundamental question: what exactly is it that we are protecting?

A huge range of creatures are at home in the air. Along with the thousands of bird species that flit from perch to perch, there are others, such as the albatross and Alpine swift, that spend much of their life aloft. Bats, mostly nocturnal fliers, often dine on the myriad insects that share their airspace. Millions of other insects, from butterflies to beetles, occupy the skies by day. Ballooning spiders are at the mercy of winds that catch long trails of web and carry them far from home. Microbes, winged seeds and spores are also all transported on the breeze, and can travel hundreds or thousands of kilometres.

If we ever consider the aerial ecosystem occupied by these creatures, we tend to think of it as one vast expanse of sky. 'The minute they take off into the air, we don't really have a mechanism in place to define that habitat type. But it's really critical,' says Christina Davy at Trent University in Ontario.

As a first step to protecting the biodiversity of airspace, Davy, along with Kevin Fraser at the University of Manitoba and Adam Ford at the University of British Columbia, put forward the idea in 2017 that we should think about aerial habitats as layers, similar to the way that marine habitats are characterised by depth. They propose three subdivisions of the troposphere, the lowest zone of the atmosphere rising to roughly 15 kilometres up. The basoaerial habitat extends from the ground up to 1 kilometre. Here human threats range from tall buildings to wind turbines and moving vehicles. The mesoaerial habitat, between 1 and 8 kilometres in altitude, is characterised by steadily decreasing temperatures and oxygen levels; the main threats here are light pollution and aircraft. In the epiaerial habitat, between 8 and 13 kilometres up, temperatures plunge towards -56 °C at mid-latitudes; its inhabitants, mainly microorganisms, require special adaptations to survive.

A better definition of habitats is only part of what's needed if 'aeroconservation' is to take off, however. For a start, we're not even really sure how big the problem is we are trying to solve. A meta-study published in 2014 put the number of birds killed in building collisions at between 365 million and 968 million a year in the US. It is estimated that 140,000 to 328,000 birds are killed annually by wind turbines and thousands by civilian aircraft. In the UK, the British Trust for Ornithology estimates that 100 million birds crash into windows annually, and in Canada, more than 50 million adult birds are thought to die each year from collisions with buildings, wind farms, communication towers and other human structures that invade the skies.

On their own, though, such numbers only say so much. 'What we have are mortality counts,' says Davy. 'We don't have the data that we need to be able to say whether [such counts equate to] 1 per cent or 100 per cent of the population.' That's because we just don't know how many creatures call the sky home.

For birds, efforts to estimate populations are well under way, aided by decades of counts, ringing schemes and newer methods such as tracking with telemetry and GPS. But for other airborne creatures, we are further in the dark. Population estimates for bats are often murky or non-existent. Some early attempts to quantify insects, meanwhile, have produced staggering numbers: more than a trillion are thought to migrate over the southern UK each year, for example.

Numbers are one thing; behaviour is another. 'We can't track three-dimensional locations of small organisms for any distance because it's too hard to put a tracking device on them,' says Robb Diehl, an ecologist with the US Geological Survey who uses radar to study migratory birds.

In the past, we have rarely looked at how aerial species

move in 3D, 'because it's easier to do in 2D', says Sergio Lambertucci at the National University of Comahue in Argentina. Tools such as accelerometers and GPS are changing that. Progress is being made in charting the behaviour of larger animals, including bats, and Lambertucci is using the technology to study several raptor species, Andean condors among them.

Until we know more, it is hard to judge the effects of our airspace incursions. But we can look at how animals in other ecosystems are affected by our activities and apply these lessons to the sky. On land, habitat fragmentation has detrimental impacts on living things, for example. In aerial habitats, this could take the shape of animals making long detours to avoid tall buildings and cities, or being lured into spending time circling light sources while travelling at night. 'What are the costs of that movement to migration duration, energetic reserves and fitness once they get to their breeding sites?' asks Ford.

## Seeing the light

Light pollution, in particular, could have a big impact in all three aerial zones. 'You can see light from outside of our atmosphere,' says Travis Longcore at the University of Southern California, who researches the impact of artificial light on biodiversity. Many studies have reported effects such as seabird chicks becoming disoriented by overhead lights on their first flight out to sea and crashing. Future research aims to find out the thresholds at which artificial light levels begin to affect the navigation, dispersal, communication and reproduction of different species, get a handle on the size of those effects and determine the size of dark refuges needed to maintain natural ecosystem processes.

Computer modelling is helping to quantify the whole-population effects of both artificial light spilling skyward and, more generally, our structural cluttering of the air. Projections for hoary bats, the species most frequently killed by wind turbines in North America, for example, suggest that lethal collisions with blades could spell serious trouble for population numbers over the next century, seeing them decline by as much as 90 per cent.

Given the ubiquity of our aerial incursions, why hasn't the idea of protecting aerial habitats come to our attention earlier? 'We are terrestrial creatures,' says Diehl. 'In our evolutionary history, we've lived off the land and, to some extent, out of the water.' He suspects that our notions of habitat are deeply ingrained and, like our science, oriented towards the landscape. We may need to step away from the biases of our senses and education to lift our gaze upwards. This offers the chance of some blue-sky thinking, says Diehl: the concept of aeroconservation is so novel that, in theory at least, solutions are limited only by our imagination.

One unknown in future efforts to protect airspace is whether it comes under the umbrella of environmental law. At the moment, even the way we define ecosystems works against aeroconservation. The International Union for Conservation of Nature, for example, recognises terrestrial, aquatic and 'other' habitats, but doesn't explicitly mention the air. This oversight extends to international policy such as the UN Convention on the Conservation of Migratory Wild Species of Animals. Neglect of airspace as habitat is problematic for creatures whose lifestyles include air travel, say Davy, Fraser and Ford.

Nevertheless, legal protection of airspace isn't without precedent. No-fly or restricted zones for drones and aircraft exist, mainly above politically or militarily sensitive zones such as the centre of Washington DC. But a no-fly zone over wildlife habitat at the Boundary Waters Canoe Area Wilderness in northern Minnesota has existed since 1948 and restrictions are in place over parts of some US national marine sanctuaries to protect marine mammals and seabirds from disturbance.

Current laws have also been invoked by campaigners. Groups including the Toronto-based Fatal Light Awareness Program have been drawing attention to bird-building collisions and rescuing injured birds for decades. 'But the biggest shift came when we found ourselves as witnesses in a court of law,' says executive director Michael Mesure.

That case, in 2010, was brought in Canada by environmental law charity Ecojustice against Cadillac Fairview, a commercial property owner and manager, after hundreds of migratory birds had died in collisions with its buildings' mirrored windows in Toronto. The judge ruled that Ontario's Environmental Protection Act and Canada's Species at Risk Act prohibit reflected daylight from building windows, because the glass creates a mirage of habitat and sky, fooling birds, with potentially fatal consequences.

As a result, bird safety is now more commonly taken into account in the planning and construction of buildings in many Canadian cities. In addition, LEED – a popular green certification scheme for buildings worldwide – is piloting the inclusion of bird-friendly architecture in its points system for 'green buildings'. Windows can be treated with special film, translucent tape or spaced wires to make them visible to birds. Avoiding positioning outdoor plants near windows may also help. But all windows reflect daylight and although the laws of Ontario say they shouldn't, in practice this isn't being enforced – and it is unclear how it could be.

Invoking building codes is no panacea. In Canada, more than 25 million birds are thought to die annually after colliding with power lines, with raptors such as owls, kestrels and eagles particularly prone. This could be tackled by placing markers on wires to make them more visible, or putting them underground. Electrocution of birds that can straddle two power lines is also a big killer, particularly of Europe's white storks. A possible solution is to increase the distance between wires.

Progress is being made here: countries such as Germany require bird protection measures to be incorporated into the design of new and upgraded power lines. With wind turbines, the UN-sponsored Migratory Flying Birds project is building protection measures into new wind energy projects along important migratory routes up through eastern Africa and the Middle East, including radar sensors that enable turbines to be shut down within minutes when a flock is approaching.

As for addressing light pollution, aeroconservationists have a natural ally: the International Dark-Sky Association. A movement founded by astronomers, its aim is to preserve some of Earth's natural darkness. In some US national parks, retrofitting has already begun to reduce upward spillage of light. Commercial lighting companies are making changes too, although progress is slow, says Longcore.

While developments are small-scale and piecemeal for now, they are no less important for the creatures concerned. These include the Blackburnian warblers that migrate through Chicago each year. Though it took time to seek expert help and navigate the bureaucracy, Meredith eventually secured an FBI-approved plan: netting put up during the migration season now protects birds from the building that had been killing them. It may temporarily restrict the spectacular views, but Meredith is convinced it is a price worth paying. 'Anything we do is going to look better than a bunch of dead birds,' she says.

# The strange creatures living far below our feet

*From microbes that have been alive since the dinosaur era to animals that get by without oxygen,* **Colin Barras** *discovers how strange underground organisms are redefining what it means to be alive – and where life will end.*

South Africa's gold mines are crawling with demons, each sporting a whip-like tail and a voracious appetite. Not that the miners are worried. These demons are barely visible to the naked eye.

They are big news for people studying life on Earth, though. 'The discovery floored me,' says Tullis Onstott, a geologist at Princeton University, whose colleague Gaetan Borgonie discovered these nematode worms swimming in the water-filled fissures of the Beatrix gold mine in 2011.

The fact is, complex organisms just shouldn't be able to live so far beneath Earth's surface. The nourishment and oxygen that animals need to survive are in short supply just tens of metres below ground, let alone 1.3 kilometres down. Noting that the worms shunned light like a mythical devil, Onstott and Borgonie's team named them *Halicephalobus mephisto*, after Mephistopheles, the personal demon of Dr Faustus.

Travelling even deeper into South Africa's crust, they found more surprises. On a trek down into TauTona, the country's deepest gold mine, they came across another species of nematode worm at 3.6 kilometres below ground – making it the deepest land animal found to date.

In fact, we now know that the depths of Earth's crust harbour isolated ecosystems whose inhabitants defy many established biological rules. There are microbes that metabolise so slowly they may be millions of years old; bacteria that survive without benefiting from the sun's energy; and animals that do what no animal should – live their entire lives without oxygen. This strange menagerie might give us insights into where life originated and where it is heading. It may even help our search for life on other worlds.

That would be an ironic twist. For most of the twentieth century, few suspected that Earth's interior could harbour any life, let alone writhing worms or scuttling bugs. Biologists were looking for signs of life on Mars long before they turned their gaze downwards. 'The prevailing view was that the deep Earth was sterile,' says Barbara Sherwood Lollar, a geologist at the University of Toronto who also studies South Africa's gold mines.

It took the nuclear arms race to overturn that orthodoxy. By the 1980s, the US had taken to burying sealed containers of radioactive waste below its nuclear processing facilities, and the Department of Energy was concerned that deep microbes, if they existed, might eat through the seals. In 1987, to ease the fears, the DoE sponsored a team to hunt for life in boreholes below the Savannah river facility in South Carolina. To general astonishment, they found bacteria and single-celled organisms called archaea 500 metres below the surface.

It didn't take long to find out that deep life was not only possible, but extremely prevalent. In 1992, John Parkes, now at the University of Cardiff, found the sediment under the Sea of Japan to be teeming with life. Even at 500 metres below the sea floor, he found 11 million microbes in every cubic centimetre of dirt.

The implications were extraordinary. Even when you consider that the heat emanating from Earth's interior would kill anything deeper than 4 kilometres below the surface, there would be enough room to house a considerable portion of the planet's life. Estimates vary from less than 1 per cent of the world's biomass to 10 per cent, with a more thorough exploration of Earth's crust needed to firm up that figure.

In the meantime, the focus has switched to answering some of the most pressing questions about the challenges facing organisms deep underground. Top of the list was the question of how they can feed in such barren regions. The microbes under the sea floor, for instance, must have originated on the seabed before being buried under sediment over thousands of years. Left with just small amounts of nutrients in the surrounding dirt, and without any new source of food, the microbes should have starved long ago. Indeed, given that the microbes are eerily still when viewed under a microscope, some sceptics argued that they were exquisitely preserved corpses of long-dead cells, rather than living organisms.

Yet that's not what Yuki Morono of the Japan Agency for Marine-Earth Science and Technology in Nankoku found in 2011. His team took cells from 460,000-year-old sediments found 220 metres below the Pacific Ocean seabed near Japan, and exposed them to a plentiful food supply labelled with stable isotopes of carbon and nitrogen. Two months later, Morono found traces of the isotopes in three-quarters of the cells. They were still alive – although you could not tell from their behaviour.

'Their lives are so slow compared with ours,' says Morono. 'It is really difficult to distinguish alive and dead cells.' The key to their survival seems to be an incredibly slow metabolism, allowing the meagre food source to be rationed for thousands of years.

## Methuselah microbes

If that lifestyle seems austere, it is nothing compared with the ecosystem discovered by Hans Røy at Aarhus University in Denmark and colleagues. Beneath the Pacific Ocean, they found active bacteria and archaea in sediments deposited 86 million years ago – 20 million years before the dinosaurs went extinct. The cells' reduced metabolism suggests each has been on a very strict diet for the entire time. Under such tight constraints, populations are very sparse, with a mere 1000 cells occupying every cubic centimetre of sediment.

Evolution may work very differently in these isolated pockets of sediment. 'If there is barely enough energy to meet the requirements of a single cell, it is suicide for that cell to divide,' says Røy. Microbes in these ancient sediments might focus their efforts on repairing their own machinery rather than bothering with the activity that most other organisms live for: reproduction.

If these ideas are right, then some of these organisms could be among the oldest creatures on the planet. 'Cells in these environments could be millions of years old for all we know,' says Katrina Edwards at the University of Southern California.

As strange as they are, the Methuselah microbes living beneath the sea are pretty conventional compared with some of the organisms found below Earth's continents. Consider one species of bacteria living down South Africa's Mponeng gold mine, whose food chain begins with the radioactive decay of minerals in the surrounding rocks.

Just reaching these microbes can be physically exhausting, says Sherwood Lollar, who helped discover the bacteria. 'You might go down in the cages with a mining shift crew at 7 a.m., and not actually reach your site until 10.30 a.m.,' she says. With temperatures and humidity almost unbearably high, the researchers have a few hours at most to collect samples from water-filled fissures in the mine's boreholes. 'Then you turn around and make the trek back up.'

At first sight, the crystalline rocks down there appear to be an even more desolate home than ocean sediment; formed deep in prehistory, they have received next to no organic matter, even in the distant past. It would seem impossible to find food down here, yet bacteria manage to eke out a meagre existence. Their secret? Uranium. As this element decays, the resulting radiation splits water molecules, releasing free hydrogen, through a process called radiolysis. The bacteria then combine the hydrogen with sulphate ions from the rock, producing enough energy to sustain life.

Powering their cells in this way, these bacteria are part of a select club of species that survive without any input from the sun. 'I would say the energy sources are all independent from photosynthetic sources,' says Li-Hung Lin at the National Taiwan University in Taipei, who led the team that discovered the bacteria.

While such discoveries extended the known boundaries of life on Earth, for a long time it seemed that deep-dwelling organisms would be limited to simple, single-celled life forms: bacteria, archaea and a few slightly more sophisticated fungi and amoebas. While they are all fascinating organisms in their own right, they are not very lively.

Then Onstott's demon worms showed that animals can live kilometres below the surface. They may be only half a millimetre long, but that still makes them hundreds of times bigger, and far more complex, than other deep dwellers. 'The diversity in the crust is greater than I ever imagined,' says Onstott.

The demon worms probably arrived in the mine relatively recently, though. Isotopic dating of the surrounding water suggests they reached the depths perhaps only 12,000 years ago, probably riding in groundwater that trickled into Earth's interior. Importantly, this water still contains oxygen from when it was last in contact with the atmosphere. Once that oxygen is used up, the worms will die, making it a fleeting stay in evolutionary terms.

But some animals have evolved to survive these suffocating conditions for the long haul, if one discovery from deep under the Mediterranean Sea is anything to go by. Found in 2010, these unusual Loricifera resemble tiny, dead houseplants – complete with pots. The 250-micrometre-long animals each have a vase-shaped armoured case, or lorica, and a straggly mess of tentacle-like projections emerging from its opening.

It isn't their appearance that makes them a showstopper for biologists, though. Antonio Pusceddu at the Marche Polytechnic University in Ancona, Italy and colleagues have found these Loricifera have evolved a unique method of metabolism that does not rely on oxygen, unlike that of all other animals. Indeed, their cells completely lack mitochondria, the organelles that power other animals. Instead, they generate energy from hydrogen sulphide using organelles called hydrogenosomes.

## No oxygen, no problem

For William Martin at the University of Dusseldorf, the Loricifera are evidence that oxygen is not the key to complex animal life. Their sluggish behaviour has caused some sceptics to question the find – just as some critics had believed the inactive underground microbes to be dead. 'Some researchers would like to see independent corroboration that the Loricifera are really alive,' he says.

If that verification comes, it would raise hopes that deep life may be far more sophisticated than anyone had imagined. That would bode well for two projects – the Census of Deep Life and the Center for Dark Energy Biosphere Investigations – that aim to catalogue underground life.

Besides giving us a better understanding of life on Earth today, the results may also give us a glimpse back in time to our early origins. At the very least, South Africa's radiolysis-powered bacteria may give us a new viewpoint on the molecular machinery that allowed life to thrive before photosynthesis reshaped the planet. Some go even further, suggesting that life itself began deep underground.

'There were tumultuous geological processes going on at the time life appeared,' says Sherwood Lollar. 'There's a strong argument to consider that life arose in a warm little fracture where it might have been protected from heavy asteroid bombardments or the lethal ultraviolet light that bathed early Earth.' It is by no means a mainstream theory: most believe hydrothermal vents in the ocean to have been the cradle of life.

But even if these fissures within the crust didn't witness the birth of the first life forms, they will almost certainly be the last refuge for organisms at the end of our planet's life. In 2012, Jack O'Malley-James at the University of St Andrews and his colleagues modelled the likely fate of life on Earth as the ageing sun makes conditions increasingly hostile. The model suggests that about a billion years from now the oceans will begin to evaporate, and the only life to survive will be microbes deep below Earth's surface, where they might hold on for another billion years. For all the grandeur of rainforests, savannahs and coral reefs, deep life is probably a more persistent feature of our planet.

The same may also be true of other worlds. 'The results from the deep biosphere are completely changing the exploration strategy for life on other planets,' says Sherwood Lollar. 'The Viking expeditions to Mars in the 1970s were looking for life on the surface. Now we know that signs of life are much more likely to be found in the subsurface.' And following the discovery of the Loricifera, there are renewed hopes for complex life forms.

For the moment, though, many eyes are gazing downwards. 'This is the last unexplored part of our planet,' says Pusceddu. 'We can expect even more exciting discoveries of animals and unicellular organisms in future.' It seems we have barely scratched the surface.

## The search for hidden treasure in alien oceans

*Deep underground isn't the only place that might show us what life could be like on other worlds. Joshua Sokol finds denizens of the sea that would also be suited to the oceans of icy moons in our solar system.*

Suddenly, out of darkness, a ghostly city of gnarled white towers looms over the submersible. As the sub approaches to scrape a sample from them, crew-member Kevin Hand spots something otherworldly: a translucent, spaceship-like creature, its iridescent cilia pulsing gently as it passes through the rover's headlights.

Hand is a planetary scientist at NASA's Jet Propulsion Lab (JPL) in Pasadena, California, and one of a select few to have visited the carbonate chimneys of the Lost City at the bottom of the Atlantic Ocean. It is the site of an extraordinary ecosystem – one that Hand suspects might be replicated on icy moons orbiting distant gas giants. 'In my head, I was saying to myself: this is what it might look like,' he says.

Jupiter's moon Europa, and Enceladus, which orbits Saturn, both have vast oceans secreted beneath their frozen outer shells. As such, many astrobiologists consider them our best bet in the search for life beyond Earth. NASA is plotting life-finding missions there. But we don't have to wait to dip our toes in extraterrestrial waters.

Having explored extreme ecosystems on our own ocean floor – places like Lost City, where life is fuelled by nothing more than the reaction between rock and water – we know what to look for. Now the race is on to spot signs of similar geochemical rumblings on Europa and Enceladus, and so discover whether we truly are alone in the solar system.

'Follow the water' has long been the mantra in the search for life, and for good reason: every known organism needs water to survive. Most prospecting has been done on Mars, but the Red Planet's water is either long gone or locked in the ground as ice. These days, even Mars buffs would struggle to deny that the best prospects for finding living extraterrestrials lie further from Earth.

It might seem odd to search for liquid water in places far from the sun's warmth. And yet it looks as if there are sloshing oceans beneath the surfaces of Europa and Enceladus, thanks to tidal flexing as a result of their eccentric orbits. As the gravity of their host planets pushes and pulls at the moons' interiors, they warm from the inside out – and that heating is enough to maintain a layer of liquid between their rocky mantles and icy crusts.

The first hints of Europa's concealed sea came from the *Voyager* probes, which explored Jupiter back in the 1970s. *Voyager 2* spotted cracks in Europa's icy surface crust, suggesting active processes below. When the *Galileo* spacecraft returned in the 1990s, it saw another clue: Jupiter's magnetic field lines were bent around Europa, indicating the presence of a secondary field. The best explanation is the presence of a global vat of electrically conductive fluid, and seawater fits the bill. We now think this ice-enclosed ocean reaches down 100 kilometres. If so, it contains enough salty water to fill Earth's ocean basins roughly twice over.

The case for a sea on Enceladus washed in more recently. In 2005, the *Cassini* probe showed that the moon leaves a distinct impression in Saturn's magnetic field, indicating the presence of something that can interact with it. That turned out to be an astrobiologist's fantasy: a plume of ice particles and water vapour shooting into space through cracks near Enceladus's south pole.

*Cassini* has since flown through these plumes several times. First its instruments revealed the presence of organic compounds. They seemed to be coming from a liquid reservoir – and the particles collected from the lowest part of the plumes were rich in salt, indicative of an ocean beneath. *Cassini* detected ammonia, too, which acts as an antifreeze to keep water flowing even at low temperatures. All the signs suggested this was a sea of liquid water, stocked with at least some of the building blocks of life.

## Treasure plumes

'After decades of scratching around Mars to find any organics at all, this was an embarrassment of riches,' says Chris McKay, an astrobiologist at NASA's Ames Research Center in Moffett Field, California.

The treasures kept coming. In March 2015, *Cassini* scientists detected silicate grains in the plumes – particles that most likely formed in reactions at hydrothermal vents. By September, measurements of how Enceladus's outer crust slips and slides had convinced them that it contains a global ocean between 26 and 31 kilometres in depth. That's a paddling pool compared with Europa's, but way deeper than Earth's oceans.

So when do we visit? NASA has already selected instruments for a new mission to Europa, set for launch in June

2022. It will feature a magnetometer to probe the ocean's saltiness and ice-penetrating radar to show where solid shell meets liquid water. It might even include a lander to fish for amino acids, the building blocks of the proteins used by every living thing on Earth.

The space agency has also invited proposals for a trip to Enceladus. One option is the *Enceladus Life Finder*, a probe that will sample plumes using instruments capable of detecting larger molecules and more accurately distinguishing between chemical signatures. Other plans have even suggested carrying samples back to Earth for analysis.

With any luck, NASA probes will be arriving at these ocean worlds by the twilight years of the 2020s. Until then we just have to sit tight, daydreaming about what fresh wonders we might find once we get there. Or do we?

In fact, there is plenty we can do in the meantime to plumb Europa and Enceladus's hidden depths. We can survey their surfaces using ground-based telescopes, gawping at the fissures where water might bubble through and leave telltale deposits from the oceans beneath. We can model the geophysics that keeps them liquid so far from the sun, and may generate conditions that could support life. And we can use the closest analogues on our own planet to guide our search.

On Earth, deep-sea vents at the boundaries between tectonic plates, where magma breaches the sea floor, have long been recognised as hotbeds for life. Around geysers of scalding, murky water – known as black smokers – bacteria feed on chemicals, and all manner of organisms make their living on those microbes. Europa or Enceladus might just draw enough energy from the tidal push and pull of their host planets to have molten interiors that can fuel similar vents. We don't know. The good news for life hunters, however, is that we're now aware of another possibility.

When we discovered the Lost City vents beneath the Atlantic in 2000, we saw that you can have a hydrothermal ecosystem with resident microbes and the occasional visit from a comb jelly – the otherworldly creature spotted by Hand during his visit – without the faintest rumble of tectonic activity.

Lost City is powered by a chemical reaction called serpentinisation. When alkaline rocks from Earth's mantle meet a more acidic ocean, they generate heat and spew out hydrogen, which in turn reacts with the carbon compounds dissolved in seawater. It is these reactions that slowly built the towers of carbonate, some 60 metres tall, that disgorge organic-rich alkaline fluids into the water and make methane for microbes to snack on.

According to Michael Russell, a geologist turned astrobiologist at JPL, Lost City is just the sort of place where life on Earth might have begun. Russell thinks that the imbalance between the alkaline fluid flooding cell-like pores inside carbonate chimneys and the relatively acidic seawater beyond created electrochemical potential that the molecular precursors of life found a way to tap. If he's right, then wherever alkaline hydrothermal vents exist life may have followed.

Astrobiologists like Hand think there is a good chance we'll find them on Europa and Enceladus. Now they are attempting to confirm their suspicions from afar. One way to do that is to look for molecules whose presence would betray ongoing serpentinisation. *Cassini*'s discovery of silicate grains in Enceladus's plumes suggests this reaction has at least happened there in the past. Recent estimates suggesting that the ocean itself is rather alkaline, which would be expected after eons of serpentinisation, add to the case. To figure out if the process is happening today, however, we want to see hydrogen. That would be important because where there are free molecules of hydrogen gas in the deep sea, there tends to be life. 'Hydrogen is chocolate-chip cookies for microbes,' says McKay.

Although *Cassini* was not built to detect molecules as large as amino acids, the probe could detect small molecules like hydrogen. During its penultimate dive through the plumes in late 2015, *Cassini*'s nose caught a clear whiff of hydrogen. But it will be tricky to distinguish between the possible sources of any hydrogen molecules they find. The trouble is that hydrogen in the plumes could either be from serpentin-isation or from water split apart in the atmosphere, after it was launched from the surface. If it turns out to be the former, it would be a big deal – the strongest indication yet that hydrothermal vents at the bottom of Enceladus's ocean are serving up good amounts of chemical fuel.

There may also be other clues in *Cassini*'s back catalogue. The probe flew through the plumes so fast that it broke apart larger compounds, and we might be able to use its detection of the fragments to reconstruct the big stuff. 'There are clearly some aromatics in some of these heavier compounds,' says Hunter Waite at the Southwest Research Institute in San Antonio, Texas. But aromatic compounds can be produced through either biological or abiotic processes, so their presence wouldn't be a smoking gun. Still, it would help us understand what kinds of carbon chemistry can flourish under the ice.

Europa is even more likely to have serpentinisation because it is much larger, meaning it boasts more rock in contact with seawater. There are no confirmed plumes to sample, though. Instead we are learning about its ocean chem-istry by peering at its surface from telescopes on Earth. In October 2015, for example, observations made with the Keck Observatory in Hawaii revealed a strange-looking substance in a region of Europa riddled with cracks. Although the chemical signature suggests it could be dirty water ice, the dirty part has so far defied identification.

Patrick Fischer of the California Institute of Technology in Pasadena, who led the analysis, says the deposits could be potassium chloride or sodium chloride. Both are normally transparent but could be rendered visible by the shower of energetic particles raining down from volcanoes on Io, Europa's explosive sister moon. If so, we could be looking at salts left behind after underground water breached the surface and then evaporated. That would suggest the ocean is seasoned not with the sulphate salts from Io, as most people expected, but with chloride – making it perhaps a third as salty as expected and therefore friendlier to life.

In 2016, Hand and his colleagues made a bigger splash with a study suggesting that Europa's ocean has a chemical balance similar to Earth's. The calculations were based on estimates that fractures in the moon's sea floor could reach as deep as 25 kilometres into the rocky interior. In that case, there would be great swathes of rock surface with which water can react to release lots of hydrogen.

But that is just one part of a cycle required for life as we know it: electron-grabbing oxidants like oxygen and electron-giving reducing agents like hydrogen have to meet and react, releasing energy that living things rely on in the form of electrons.

Europa has no atmosphere from which to cycle oxygen, as Earth does, but we know that radiation from Jupiter produces oxidising chemicals on its surface. To arrive at their conclusions about Europa's sea, Hand and his colleagues assumed that these oxidants are being cycled from surface to sea.

## Life, cycled

That's a big assumption. 'If you mix the subsurface and the surface, then you get a chemical cycle that life could take advantage of,' says Britney Schmidt, an astrobiologist at the Georgia Institute of Technology in Atlanta. If not, life is unlikely. And it's not yet clear whether that cycling happens on Europa, never mind Enceladus, where the radiation from Saturn is weaker, leaving fewer oxidants on its surface.

To find out, Schmidt has drilled through Antarctic sea ice and deployed a robotic submarine to study the underside, where fresh ice is constantly forming and melting. 'If we can figure out how the ice and ocean system works here on Earth, then we can extrapolate back to Europa,' she says. Only then will we know if its vast ocean gets enough oxidants to create the ratio of elements for life.

It is possible, of course, that life elsewhere follows a different rulebook, that it is made from a different set of biochemical building blocks. So what should we be looking for if not organic molecules and amino acids? It is a question that astrobiologists contemplate, but it can probably be answered only by finding alien life forms.

Maybe we never will. Maybe we really are alone in the solar system. If we can detect something akin to deep-sea alkaline vents on faraway moons, however, the odds of finding extraterrestrials would be slashed.

We might also have to entertain the prospect that similarly life-friendly conditions are lurking beneath the shells of other icy worlds: moons like Ganymede, Mimas and Ceres. In fact, given how common we now know them to be, oceans concealed by frozen crusts could be the default condition for life – in which case our blue planet, with its peculiar open oceans, is the outlier.

## The electric life forms that live on pure energy

*Unlike any other life on Earth, these extraordinary bacteria use energy in its purest form – they eat and breathe electrons – and they are everywhere, says* **Catherine Brahic.**

Stick an electrode in the ground, pump electrons down it, and they will come: living cells that eat electricity. We have known bacteria to survive on a variety of energy sources, but none as weird as this. Think of Frankenstein's monster, brought to life by galvanic energy, except these 'electric bacteria' are very real and are popping up all over the place.

Unlike any other living thing on Earth, electric bacteria use energy in its purest form – naked electricity in the shape of electrons harvested from rocks and metals. We already knew about two types, *Shewanella* and *Geobacter*. Now, biologists are showing that they can entice many more out of rocks and marine mud by tempting them with a bit of electrical juice. Experiments growing bacteria on battery electrodes demonstrate that these novel, mind-boggling forms of life are essentially eating and excreting electricity.

That should not come as a complete surprise, says Kenneth Nealson at the University of Southern California, Los Angeles. We know that life, when you boil it right down, is a flow of electrons: 'You eat sugars that have excess electrons, and you breathe in oxygen that willingly takes them.' Our cells break down the sugars, and the electrons flow through them in a complex set of chemical reactions until they are passed on to electron-hungry oxygen.

In the process, cells make ATP, a molecule that acts as an energy storage unit for almost all living things. Moving electrons around is a key part of making ATP. 'Life's very clever,' says Nealson. 'It figures out how to suck electrons out of everything we eat and keep them under control.' In most living things, the body packages the electrons up into molecules that can safely carry them through the cells until they are dumped on to oxygen.

'That's the way we make all our energy and it's the same for every organism on this planet,' says Nealson. 'Electrons must flow in order for energy to be gained. This is why when someone suffocates another person they are dead within minutes. You have stopped the supply of oxygen, so the electrons can no longer flow.'

The discovery of electric bacteria shows that some very basic forms of life can do away with sugary middlemen and handle the energy in its purest form – electrons, harvested from the surface of minerals. 'It is truly foreign, you know,' says Nealson. 'In a sense, alien.'

Nealson's team is one of a handful that is now growing these bacteria directly on electrodes, keeping them alive with electricity and nothing else – neither sugars nor any other kind of nutrient. The highly dangerous equivalent in humans, he says, would be for us to power up by shoving our fingers in a DC electrical socket.

To grow these bacteria, the team collects sediment from the seabed, brings it back to the lab, and inserts electrodes into it. First they measure the natural voltage across the sediment, before applying a slightly different one. A slightly higher voltage offers an excess of electrons; a slightly lower voltage means the electrode will readily accept electrons from anything willing to pass them off. Bugs in the sediments can either 'eat' electrons from the higher voltage, or 'breathe' electrons on to the lower-voltage electrode, generating a current. That current is picked up by the researchers as a signal of the type of life they have captured. 'Basically, the idea is to take sediment, stick electrodes inside and then ask "OK, who likes this?"' says Nealson.

## Shocking breath

At the 2014 Goldschmidt geoscience conference in Sacramento, California, Shiue-lin Li of Nealson's lab presented results of experiments growing electricity breathers in sediment collected from Santa Catalina harbour in California. Yamini Jangir, also from the University of Southern California, presented separate experiments that grew electricity breathers collected from a well in Death Valley in the Mojave Desert in California.

Over at the University of Minnesota in St Paul, Daniel Bond and his colleagues have published experiments showing that they could grow a type of bacteria that harvested electrons from an iron electrode. That research, says Jangir's supervisor Moh El-Naggar, may be the most convincing example we have so far of electricity eaters grown on a supply of electrons with no added food.

Nealson is particularly excited that Rowe has found so many types of electric bacteria, all very different to one another, and none of them anything like *Shewanella* or *Geobacter*. 'This is huge. What it means is that there's a whole part of the microbial world that we don't know about.'

Discovering this hidden biosphere is precisely why Jangir and El-Naggar want to cultivate electric bacteria. 'We're using electrodes to mimic their interactions,' says El-Naggar. 'Culturing the "unculturables", if you will.' The researchers plan to install a battery inside a gold mine in South Dakota to see what they can find living down there.

NASA is also interested in things that live deep underground because such organisms often survive on very little energy and they may suggest modes of life in other parts of the solar system.

Electric bacteria could have practical uses here on Earth, however, such as creating biomachines that do useful things like clean up sewage or contaminated groundwater while drawing their own power from their surroundings. Nealson calls them self-powered useful devices, or SPUDs.

Practicality aside, another exciting prospect is to use electric bacteria to probe fundamental questions about life, such as what is the bare minimum of energy needed to maintain life.

For that we need the next stage of experiments, says Yuri Gorby, a microbiologist at the Rensselaer Polytechnic Institute in Troy, New York: bacteria should be grown not on a single electrode but between two. These bacteria would effectively eat electrons from one electrode, use them as a source of energy, and discard them on to the other electrode.

Gorby believes bacterial cells that both eat and breathe electrons will soon be discovered. 'An electric bacterium grown between two electrodes could maintain itself virtually forever,' says Gorby. 'If nothing is going to eat it or destroy it then, theoretically, we should be able to maintain that organism indefinitely.'

It may also be possible to vary the voltage applied to the electrodes, putting the energetic squeeze on cells to the point at which they are just doing the absolute minimum to stay alive. In this state, the cells may not be able to reproduce or grow, but they would still be able to run repairs on cell machinery. 'For them, the work that energy does would be maintaining life – maintaining viability,' says Gorby.

How much juice do you need to keep a living electric bacterium going? Answer that question, and you've answered one of the most fundamental existential questions there is.

# 4 The Threadbare Fabric of Reality

How sure are you that the world is real? It feels solid, but peer closely enough and discrete atoms dissolve into a haze of wave-particles. Squint even closer, and these clouds of probability unravel further into abstract fields with properties that seem impossible. Does a universe-wide quantum substrate set every atom in existence into motion? Why do we need imaginary numbers to describe real things? And how do entangled photons whisper secrets to each other at thousands of times the speed of light?

In this chapter, we inspect the ethereal foundations of reality – you'll be amazed the universe hasn't fallen down already.

# Why does the universe even exist?

*Perhaps the biggest question of all time is why anything exists.* **Amanda Gefter** *asks if there is a reason why there is something rather than nothing.*

As Douglas Adams once wrote: 'The universe is big. Really big.' And yet if our theory of the big bang is right, the universe was once a lot smaller. Indeed, at one point it was non-existent. Around 13.7 billion years ago time and space spontaneously sprang from the void. How did that happen?

Or to put it another way: why does anything exist at all? It's a big question, perhaps the biggest. The idea that the universe simply appeared out of nothing is difficult enough; trying to conceive of nothingness is perhaps even harder.

It is also a very reasonable question to ask from a scientific perspective. After all, some basic physics suggests that you and the rest of the universe are overwhelmingly unlikely to exist. The second law of thermodynamics, that most existentially resonant of physical laws, says that disorder, or entropy, always tends to increase. Entropy measures the number of ways you can rearrange a system's components without changing its overall appearance. The molecules in a hot gas, for example, can be arranged in many different ways to create the same overall temperature and pressure, making the gas a high-entropy system. In contrast, you can't rearrange the molecules of a living thing very much without turning it into a non-living thing, so you are a low-entropy system.

By the same logic, nothingness is the highest entropy state around – you can shuffle it around all you want and it still looks like nothing.

Given this law, it is hard to see how nothing could ever be turned into something, let alone something as big as a universe. But entropy is only part of the story. The other consideration is symmetry – a quality that appears to exert profound influence on the physical universe wherever it crops up. Nothingness is very symmetrical indeed. 'There's no telling one part from another, so it has total symmetry,' says physicist Frank Wilczek of the Massachusetts Institute of Technology.

And as physicists have learned over the past few decades, symmetries are made to be broken. Wilczek's own speciality is quantum chromodynamics, the theory that describes how quarks behave deep within atomic nuclei. It tells us that nothingness is a precarious state of affairs. 'You can form a state that has no quarks and antiquarks in it, and it's totally unstable,' says Wilczek. 'It spontaneously starts producing quark-antiquark pairs.' The perfect symmetry of nothingness is broken. That leads to an unexpected conclusion, says Victor Stenger, a physicist at the University of Colorado in Boulder: despite entropy, 'something is the more natural state than nothing'.

'According to quantum theory, there is no state of "emptiness",' agrees Frank Close of the University of Oxford. Emptiness would have precisely zero energy, far too exacting a requirement for the uncertain quantum world. Instead, a vacuum is actually filled with a roiling broth of particles that pop in and out of existence. In that sense this book, you, me, the moon and everything else in our universe are just excitations of the quantum vacuum.

## Before the big bang

Might something similar account for the origin of the universe itself? Quite plausibly, says Wilczek. 'There is no barrier between nothing and a rich universe full of matter,' he says. This, of course, raises the question of what came before the big bang, and how long it lasted. Unfortunately, at this point basic ideas begin to fail us; the concept 'before' becomes meaningless. In the words of Stephen Hawking, it's like asking what is north of the North Pole.

Even so, there is an even more mind-blowing consequence of the idea that something can come from nothing: perhaps nothingness itself cannot exist.

Here's why. Quantum uncertainty allows a trade-off between time and energy, so something that lasts a long time must have little energy. To explain how our universe has lasted for the billions of years that it has taken galaxies to form, solar systems to coalesce and life to evolve into bipeds who ask how something came from nothing, its total energy must be extraordinarily low.

That fits with the generally accepted view of the universe's early moments, which sees space–time undergoing a brief burst of expansion immediately after the big bang. This heady period, known as inflation, flooded the universe with energy. But according to Einstein's general theory of relativity, more space–time also means more gravity. Gravity's attractive pull represents negative energy that can cancel out inflation's positive energy – essentially constructing a cosmos for nothing. 'I like to say that the universe is the ultimate free lunch,' says Alan Guth, a cosmologist at MIT who came up with the inflation theory 30 years ago.

Physicists used to worry that creating something from nothing would violate all sorts of physical laws, such as the conservation of energy. But if there is zero overall energy to conserve, the problem evaporates – and a universe that simply popped out of nothing becomes not just plausible, but probable. 'Maybe a better way of saying it is that something *is* nothing,' says Guth.

None of this really gets us off the hook, however. Our understanding of creation relies on the validity of the laws of physics, particularly quantum uncertainty. But that implies that the laws of physics were somehow encoded into the fabric of our universe before it existed. How can physical laws exist outside of space and time and without a cause of their own? Or, to put it another way, why is there something rather than nothing?

## The quantum battle for the basis of reality

*Reality, relativity, causality or free will? Take quantum theory at face value and at least one of them is an illusion. Figuring out which is no easy task, writes* **Michael Brooks**.

At first, it looks like an ordinary mirror. But it's not. It is 'half-silvered'. Half of the light that hits it is reflected. The other half passes straight through. This is not in itself extraordinary. Any time you look out of a window and see the room you are sitting in partially reflected, you're seeing a similar effect. Special beam-splitting mirrors are part of any teleprompter, and you can buy them online without breaking the bank if you really want to.

It's what they do to the individual photons of light that's the strange thing. Peer too closely, and these looking glasses might destroy your very perception of reality. They could leave you unsure where or even who you are, and make you question whether you exist at all. They might even skew your notions of cause and effect so much that they leave you wondering whether you, rather than the mirror, are to blame for all this. The question is whether science gets any more profound than what happens at a half-silvered mirror. 'I don't think it does,' says Terry Rudolph, a physicist at Imperial College London.The culprit, as usual when we find ourselves assailed by doubt and racked with existential fear, is quantum theory. Quantum theory is our best stab yet at delivering a picture of how material reality works at the smallest scales, and its predictions have been confirmed time and time again by experiment. It's just that the reality it describes seems to bear little relation to . . . well, reality.

For a start, quantum reality is unpleasantly random. Take the atoms of something as apparently real as us. According to quantum theory, when in isolation they are never definitely in any one place. There is only ever a certain probability of finding an atom at point X, a different probability of finding it at Y, and yet another of finding it at Z. As long as you don't ask where it is, an atom exists in a 'superposition' of all the possible places it might be. Ask the question – make a measurement – and the atom will reveal itself to be some-where, but you won't necessarily be able to predict where.

That weirdness reaches its apogee at the half-silvered mirror. Let light hit the mirror in the right way, and it is not just the light beam that is split, but individual photons. They become in effect two photons. One of them passes through the mirror, and the other is reflected.

Each of these photons has particular properties, for example spin – a quantum-mechanical quantity that can be envisaged rather like a rotation in space. But something very odd happens when you decide to measure these spins one after the other. You can do this over and over again, each time measuring the two spins relative to different things – the lab floor, the direction of the prevailing wind outside, the direction in which a fly is walking across the ceiling above. After a while, a chill runs down your spine. A pattern emerges: each time, the outcome of the second measurement depends on how you chose to make the first measurement.

This is something we cannot explain with normal, classical conceptions of reality. It is entanglement: the ability of quantum objects that were once related to apparently influence each other's properties when subsequently separated, even by a long way. Spooky action at a distance, in Einstein's words.

It is tempting to seek succour in some 'normal' physical explanation for this, as Einstein did, and so maintain our standard perceptions of reality. There must be some undetected influence that flies between the two photons. Something physical must pass from one to inform the other of the information that has been extracted.

Whatever form that influence might take – a photon, some other exchanged particle or perhaps a type of wave? – a good guess is that it will not travel faster than the speed of light. Thanks to Einstein's relativity, that is always seen as a kind of fundamental speed limit to any kind of usable information flying through the universe. Having that limit prevents all sorts of unpleasant consequences. 'We would have weird situations, weird violations of causality, if there were super-luminal signalling,' says Rudolph. Any faster-than-light channel might also be open to hijacking for nefarious purposes: you could use it to transmit information backwards in time. Allow violations of relativistic causality, and we could all be lottery millionaires.

Hidden physical influences of the less outlandish sort, which obey relativity, can be tested for relatively easily. First you separate two entangled photons by a huge distance. The second photon is sent away – to the International Space Station (ISS), say – with instructions to carry out a measurement at a precise time. An instant before that measurement occurs, you measure the first photon. Time it right, and there is not enough time for any influence to travel between the two, even at the speed of light.

## Shaken and stirred

Nobody has yet done the ISS test, but we have done similar things many times on Earth. Each time, when the report of the second measurement comes back, the weird influence has still been felt. The second photon responds to measurements as if it were aware of what happened to the first. Experiments performed by Nicolas Gisin and his colleagues at the University of Geneva in 2008 showed that any spooky influences travelling 18 kilometres through a fibre-optic network must be travelling at a minimum of 10,000 times the speed of light. The experiments have also been done over hundreds of kilometres in free air with similar results, and in 2017 China beamed entangled photons from a satellite 1200 kilometres to ground stations on Earth.

So where does this leave us? Perhaps shaken by these strange tales of unexplained correlations, you might also be stirred to accept another explanation, however far-fetched it seems at first. Relativity only forbids an influence propagating above light speed when it carries information. So what if some weird phenomenon unknown to physicists could break relativity, connect two entangled particles, while being information-free?

We have even less idea what that sort of influence might look like. Chances are it doesn't matter: since 2012, this escape route back to normality has also been blocked off. Together with Gisin and others, Jean-Daniel Bancal at the University of Geneva worked through what would happen within a network of four senders and receivers that could synchronise their measurements of entangled photons. In this theoretical set-up, influences could travel through space–time at whatever speed they liked, just as long as they contained no information.

And it failed to reproduce reality. There was no way any physical mechanism of any stamp could produce the quantum correlations seen in experiments unless hidden influences within the network could also send information at above light speed. If we trust in relativity, that leaves us with a problem. 'It pushes the weirdness further than we thought,' says Bancal. 'You go to try to find the causes of these correlations, but somehow they're just not there.'

Gisin is even more forthright in his conclusion. For him, it means that the dimensions of reality we move in cannot possibly contain the explanation for a more fundamental quantum reality. 'There is no story in space and time that tells us how the correlations happen,' he says. 'There must exist some reality outside of space–time.'

Unless there is something fundamental that we have wrong. Violations of relativity are frowned upon because they violate our ideas of causality. We humans are suckers for causal order, looking back in time to trace the cause of any event. Even more basically, we are determined determinists, blithely assuming that every event actually has a cause. That seems to work reasonably well in our large-scale everyday world, but when it comes to the nitty-gritty of the underlying quantum reality, can we be so sure?

Theorist Caslav Brukner and his colleagues at the University of Vienna recently set out to investigate whether quantum systems are subject, in theory, to the same causal laws as the rest of us. They started off from the classic situation in which two independent observers, Alice and Bob, make a measurement on a photon. The twist Brukner and his team added was quantum uncertainty, a principle that fundamentally constrains the amount of information you can extract from a quantum system – including information about time.

Brukner describes the scenario they uncovered as akin to having Alice walk into a room and find a message written by Bob. She erases it, and writes a reply – then Bob comes in to write the original message that Alice has just replied to. In effect, just as quantum particles can be in two or more places at once, so seemingly can this particle be in two or more moments at once. The system can be simultaneously in the states 'Alice came into the room before Bob' and 'Bob came into the room before Alice'. 'We cannot say whether Alice's measurement is ahead of Bob's measurement, or the other way round,' says Brukner.

Brukner is already thinking about ways to test the results of these theoretical calculations in experiment, but it won't be easy, he says. Given the delicate nature of quantum states, any attempt to measure a quantum-mechanical superposition of causal orders destroys that superposition, collapsing it into a definite causal order.

Even without experimental confirmation, though, he thinks the conclusion is clear. 'Causal order is not a fundamental property of nature,' he says. Causality is only restored when the parameters of the experiment are tweaked to make the particle behave more like familiar, classical particles. That leads him in the same sort of direction as Gisin. We live in space–time, and experience causal order within it, yet causal order is not apparently fundamental to quantum theory. If we accept quantum theory as the most fundamental description of reality that we have, it means that space–time itself is not fundamental, but emerges from a deeper, currently inscrutable quantum reality.

## Spontaneous universe

If we accept quantum theory, that is. All the havoc quantum theory wreaks with cherished notions of reality, relativity and causality raises a natural question: is quantum theory itself the problem? For all its successes, perhaps all that randomness, uncertainty and spooky influence is just because quantum mechanics is incomplete. As currently formulated, at least, it might simply not supply all the information we need to explain why things are as they are. An analogy might be made with the laws of thermodynamics. They provide a foolproof, high-level description of how things work – heat always passes from the hotter to the cooler – while saying nothing about the under-lying dynamics of individual atoms that makes that happen.

To investigate this possibility, Roger Colbeck and Renato Renner of the Swiss Federal Institute of Technology (ETH) in Zurich have taken a look at what would happen in those classic Alice-and-Bob-type experiments if an underlying theory were to provide an additional, arbitrary amount of information about the correlations between two entangled particles. Do the outcomes of the measurements look any less random and unpredictable?

The short answer is no. In any situation where both Alice and Bob can independently choose the type of measurement they make on their particle, additional information doesn't make their predictions of what will happen in experiments any more accurate than if they use quantum theory. The mysterious unpredictability of quantum mechanics has nothing to do with incomplete information, it seems.

'The randomness is intrinsic,' says Colbeck. Deep down,

the universe is spontaneous. Fundamentally, there is no reason why a quantum particle has the properties it does: there is no hidden influence, no cast-iron cause and effect, no missing information. Things are as they are; there is no explanation.

'Some people find this very depressing,' Colbeck says. So depressing, in fact, that it leads them to question an even more fundamental assumption about reality and our relation to it. It lies in a little clause in the way most investigations of quantum reality and quantum measurements, including Colbeck and Renner's, are set up. Let's go back to the first experiment, the one with the photons at the half-silvered mirror. To measure the direction of the photons' spins, you must first choose something to measure them relative to – the lab, the wind, the fly on the ceiling. Your choice influences the outcome of the measurement. But what if it is not actually your choice? What if something else were forcing your hand, making you perform the experiments such that the correlations always appear?

## In thrall to ourselves

This takes us into the domain of human free will, a slippery territory where philosophers are usually more abundant than physicists. It sounds vaguely loopy, yet some serious physicists think that a lack of free will – that we are participants in something of a cosmic puppet show – might be the best way to save us from all the weirdness and loss of relativity and causality implied by quantum correlations.

Nobel laureate Gerard 't Hooft of the University of Utrecht, for example, is one who finds the idea of quantum correlations that defy notions of space and time 'difficult to buy'. He thinks the answer might instead lie in an extreme form of determinism in which human minds are set on a trajectory of choices, such as what to make a quantum measurement relative to, from which they are powerless to deviate.

Others are less impressed. 'Invoking conspiratorial correlations among all the brains, measuring instruments, and subatomic particles in the universe to make it "look like" quantum mechanics is true is vastly stranger than the thing it's supposedly trying to explain,' says Scott Aaronson, a quantum physicist at the Massachusetts Institute of Technology. In essence, he says, there is little difference between invoking something like that and invoking a superhuman deity.

Terry Rudolph doesn't have an answer – no one does. But he reckons the problem is that we are still hopelessly anthropocentric. The growing disconnection between our experience of the world and the results of quantum experiments, he says, are simply a modern version of the ever-more complex epicycles that Ptolemy and those who followed him used to explain the motions of the heavenly bodies. The problem back then was that we could only see the planets as revolving around Earth; it took Copernicus to turn things around, and suddenly all was plain and simple.

Perhaps we have constructed theories such as relativity and quantum theory with a similarly limited view, in thrall this time to a sense of space and time that might not exist beyond ourselves. 'We think time and position and so on are important variables for describing the world because we evolved to perceive them,' says Rudolph. 'But whatever is going on down there doesn't seem to worry about them at all.'

So there you have it. When the light shines on that the half-silvered mirror, what we see is hardly a reflection of the world as we would like to know it. Reality, relativity, causality, free will, space and time: they can't all be right. But which ones are wrong?

# The particle that blew up the universe

*It gives everything mass, and now some people think the Higgs boson sparked the stupendous split-second inflation that made the cosmos we see today.* **Jon Cartwright** *investigates how a subatomic fuse may have ignited the big bang.*

They say it started with a bang, but in truth it misfired. The universe began as a hot speck of energy and, for an instant, remained just that. Then it blew up: from this initial seed, trillions upon trillions of times smaller than an atom, everything suddenly ballooned into the gargantuan proportions of a Tic Tac. In a mere fraction of a second, the universe expanded by nearly as many orders of magnitude as it would in the following 13.8 billion years.

Believe it or not, this burst of cosmological inflation, followed by a slower, tamer expansion, is the most sensible way to explain how the universe looks today. But there's something missing: what did the inflating?

The answer could be everywhere, and right under our noses. When a long-sought particle finally appeared in 2012, it seemed to close a chapter in physics without giving any clue about what happens next. Read between the lines, though, as some theorists recently have, and you see that the famous Higgs boson – the particle that gives mass, or inertia, to all other particles – might have an explosive secret. 'If the Higgs gives inertia to particles,' says Juan García-Bellido at the Autonomous University of Madrid, 'can it give inertia to the entire universe?'

Inflation wasn't always in our cosmological story. For a long time, theorists assumed that the universe expanded steadily from the start, with no sudden burst. This was driven by the natural desire of energy to spread itself in all directions.

Yet something about this picture didn't quite ring true. Look at large enough scales, and the stars, galaxies and other structures in the universe don't appear to be scattered randomly; even matter at opposite ends of the universe seems to be distributed in the same pattern of webs and clusters wherever you look, almost as if some interaction evened them out as the universe expanded. But that's impossible. Something would have had to travel faster than the speed of light between these distant points – a serious physics no-no.

## Explosive urge

Inflation, first proposed by Alan Guth at the Massachusetts Institute of Technology and others in the early 1980s, offers a way out. The idea is that a minuscule fraction of the primordial cosmos ballooned exponentially in the blink of an eye. Tiny, short-lived quantum fluctuations that are always bubbling away in space–time got caught in the whirlwind expansion and amplified, becoming the seeds of the stars and galaxies we see today. The patterns didn't evolve – they were there from the start.

There are plenty of other reasons to like inflation, but they don't change the fact that we haven't got a clue how it happened. Cosmologists suppose there must be an 'inflaton', an energy field with dynamite properties. But what and where is it?

We have an idea about what to look for: the inflaton must be a scalar field. This is just a mathematical way of describing a field that acts in all directions but whose strength can change over space and time. One way to think about it is like a weather map of air pressure. Air pressure varies depending on the location and day of the forecast, but unlike wind strength, say, it is directionless.

What we are searching for, then, is an invisible fluid-like substance, one that suffuses all of space and has the potential to influence everything in it. Or at least it did. The inflaton field must have generated something akin to extreme anti-gravity – a cosmic urge that blew up the fabric of space–time – but then quickly lost its impetus, to the point at which its influence essentially disappeared and normal expansion resumed.

In principle, there is nothing stopping us from tracking an invisible energy field that has lain low for 13.8 billion years. Particle physicists can isolate a little pocket of a field, otherwise known as a particle, by smashing other particles together to generate a momentary flash of energy. We discovered some of the most elusive fundamental particles, such as the quarks, in this way. But those particles are not associated with scalar fields. And in the decades following the proposal of inflation, our best particle colliders failed to find anything that was.

Then, in 2012, a fundamental scalar particle finally showed up: the Higgs boson. Discovered at the Large Hadron Collider (LHC) at CERN on the France–Switzerland border, the Higgs had long been predicted as the particle that endows all others with mass. Finding it was a momentous triumph.

As the world celebrated, however, a handful of theorists saw the arrival of the Higgs in a different light. Two of these were Fedor Bezrukov and Mikhail Shaposhnikov at the Swiss Federal Institute of Technology in Lausanne (EPFL). Having anticipated the Higgs discovery for several years, they had begun to consider what attributes it might have besides the gift of mass-giving.

On the face of it, the Higgs and the inflaton are different in a crucial way. Although both are scalar fields, unlike the inflaton, some of the Higgs field remains when it falls into its lowest energy state. It is precisely this sticky residue that manifests as the property of mass for other fundamental particles. But that is in today's universe. Bezrukov and Shaposhnikov realised that it is possible to tweak the properties of the Higgs field so that in the moment following the big bang, it could have mustered enough force to flood the still-minuscule cosmos with inflationary gusto.

They fiddled with the Higgs's 'potential curve' – essentially the energy a particle needs in order to have a certain effect, such as bestowing mass on other particles. Picture this as a ball on a steep-sided hill. For most particles, when the background energy is low, the ball comes to rest in the valley. The particle's location determines its effect, and right in the middle of the valley the effect is 'zero', meaning the particle is essentially switched off.

The Higgs is special, however, in that its potential curve is shaped not like a typical valley, but like the bottom of a champagne bottle, with a bump in the middle. Given that it would take energy to push the ball up that central bump, when background energy is low, the Higgs comes to rest in the valley to one side, where it turns 'on'. This is how the Higgs has the effect of giving mass to other particles, even when its field has no external energy to fuel it.

Bezrukov and Shaposhnikov spotted that there was nothing in known Higgs behaviour to stop them from adjusting the sides of its potential curve. What if, at some point high up on the curve, those steep sides flattened out somewhat? If the ball was hoisted up there for a brief time, the Higgs could sit in a supercharged 'on' state, where it would flood the universe with extreme antigravity, enough to drive apart space–time itself.

## Unexpected accomplice

True, it would require a hefty shot of energy to scale the sides in the first place. But there was an awful lot of background energy around at time-zero. 'The Higgs can make the universe expand,' says Bezrukov, now at the University of Manchester. 'It could be the inflaton.'

Finding out for sure would require a test of the particle's interaction with gravity: if the Higgs interacts strongly with gravity, then the sides of the potential curve might be flattened out as the researchers propose. Unfortunately, gravity is too weak on Earth for that to be measured at the LHC, so collider data alone can't tell us whether the Higgs potential has inflaton-capable flattened sides.

For García-Bellido, the tidiness of the Higgs explaining both the origin of mass and inflation was too hard to resist. But the more he thought about it, the less tidy things seemed. In 2011, working with Shaposhnikov and others, García-Bellido realised that the mathematical tweaking of the Higgs potential created an imbalance in the underlying equations, one that could only be remedied by a second scalar particle. This was a surprise, but not necessarily an unwelcome one.

Since the late 1990s, astronomers have known that the universe's current expansion is accelerating. They suggested that some unknown source of energy is behind the acceleration, and modern observations indicate that it must account for more than two-thirds of all the energy in the universe. The only problem is that no one knows what this so-called dark energy is.

Handy, then, that the new particle inferred by García-Bellido and his colleagues could have just what it takes to solve this mystery too. It wouldn't be nearly as burly as the Higgs, but according to the team's calculations, its field would be present in small quantities throughout the lifetime of the universe, providing just the right boost to expansion. 'That's the beauty of this model,' says García-Bellido. 'First it solves inflation; second, the accelerated expansion of the universe. It's extremely economical.'

The name of the new particle, the dilaton, reflects its close entwinement with Higgs physics. Specifically, it would prevent the Higgs's mass from 'dilating' too much – useful because without it we don't have much of a clue why the Higgs mass has the value it does. So although the dilaton itself would be massless, it would be an influential background operator, fixing the mass of the Higgs and, by extension, all other fundamental particles. Dark energy would be its biggest footprint in the universe.

Bold claims indeed. Alas, not everyone is won over. Veronica Sanz at the University of Sussex thinks that the Higgs as mass-giver and inflaton is too contrived. Instead, she backs the possibility that the Higgs and the inflaton are part of a whole new family of scalar particles that we have barely begun to uncover. All other known particles reside in families, she says, so why should the Higgs be the only scalar particle?

If Sanz is right, the Higgs would be a mere spectator in the early universe, but it would be influenced by its sibling, the inflaton, and that would show up in Higgs data from the LHC. Sanz says she prefers that idea because it is easier to test with current colliders than that proposed by García-Bellido and his colleagues. 'In cosmology, there's always a plethora of ideas that are hard to tell apart experimentally,' she adds. 'I don't like that.'

It's a fair point: the LHC alone can't rule out the Higgs-dilaton model, because the dilaton would hardly interact with other particles. But to see if the Higgs has what it takes to put the bang into the big bang, García-Bellido is not relying on the LHC. Instead, he will stare into the distance to study the after-glow of the big bang, the cosmic microwave background.

The smoking gun for an inflationary Higgs would be a

particular twist in the polarisation of this ancient light. The presence of a dilaton field would be trickier to spot, but not impossible. García-Bellido thinks it should have left a mark in any gravitational waves that imprinted themselves on the background after the tumult of inflation. Broadly speaking, that means making precise measurements of differences in the levels of microwaves coming from various directions in space.

Currently, the best picture we have of the microwave background is that recorded by the European Space Agency's Planck spacecraft in 2013. There wasn't quite enough detail for García-Bellido's purposes, but a raft of new instruments ought to do the trick. They include the Simons Observatory, under development in the Atacama Desert in Chile, Japan's forthcoming LiteBIRD satellite and the latest BICEP/Keck telescope at the South Pole.

We will have to be patient. The Simons Observatory will only begin studying the heavens in the next few years, while LiteBIRD is not due for launch until the 2020s. For García-Bellido, however, that is not long to wait to solve two of the greatest mysteries of cosmology in one fell swoop. 'We're on the verge of a breakthrough,' he says.

## The geometry that could reveal the true nature of space–time

*The discovery of an exquisite geometric structure is forcing a radical rethink of reality.* **Anil Ananthaswamy** *discovers it could clear the way to a quantum theory of gravity.*

For years after the physicist Richard Feynman died, his 1970s yellow-and-tan Dodge minivan lay rusting in a garage near Pasadena, California. When it was restored in 2012, special effort was made to repaint the giant doodles that adorned its bodywork. They don't look like much – simple combinations of straight lines, loops and squiggles. But it is no exaggeration to say these Feynman diagrams revolutionised particle physics. Without them, we might never have built the standard model of particles and forces, or discovered the Higgs boson.

Now we could be on the cusp of a second, even more far-reaching transformation. Because even as Feynman's revolution seems to be fizzling out, physicists are discovering hints of deeper geometric truths. If glimpses of exquisite mathematical structures that exist in dimensions beyond the familiar few can be substantiated, they would seem to point the way to a better understanding not just of how particles interact, but of the nature of reality itself.

It was a hard road that led to the standard model, this monumental theoretical construct that describes all the particles of the quantum world and the forces that act on them, except for gravity. The starting point came in the 1930s and early 1940s, when physicists investigating quantum electrodynamics, the theory of how charged particles and electromagnetic fields interact, embarked on calculations of 'scattering amplitudes' – the probabilities of different outcomes in a given particle interaction. But the calculations proved maddeningly difficult. For a while they seemed impossible.

Then along came Feynman. In 1949, he showed an intuitive way to tackle the calculations, using doodles that could literally be drawn on cocktail napkins. Take, for example, the interaction of two electrons. The electrons are depicted by two straight lines that are approaching each other. But before the lines meet, the electrons interact by exchanging a 'virtual' photon, drawn as a squiggly line, causing the two straight lines to move apart. The two electrons have repelled each other.

This is the simplest and most likely interaction. But for a full picture, you have to come up with all possible Feynman diagrams a given interaction could have, capturing all the different ways in which the particles can influence each other. One of the electrons might emit and absorb a virtual photon, for instance, creating a squiggly loop, which can interact with itself to generate more loops. The basic procedure is that you turn each possible diagram into an algebraic formula and work them all out to get the scattering amplitude.

The more virtual particles, the more complicated the calculations. So why invoke virtual particles at all? It does seem strange given that they are not real particles. A real particle is essentially a consistent ripple in an energy field, one that persists over time. But when real particles interact, they can cause temporary ripples in underlying quantum fields, such as an electromagnetic field. These are called virtual particles, and they are used in Feynman diagrams for several reasons.

The first is that dealing with them rather than with fields makes the maths more manageable. The other great advantage is that they help physicists visualise everything as the well-defined interactions between point-like particles, as opposed to the hazy goings-on between particles and fields. This fits nicely with the intuitive principle of locality, which holds that only things in the same spot in space and time can interact. Finally, the technique also helps enforce the principle of unitarity, which says that the probability of all outcomes should add up to 1.

## Sticky like gluons

Feynman diagrams worked beautifully when applied to photons and electrons, and became a staple of physics, being used to predict the outcome of experiments to astonishing precision. But once physicists started to tackle quantum chromo-dynamics, the theory of interactions involving quarks and gluons – the basic components of the protons and neutrons at the heart of atoms – things got sticky. There were so many virtual particles, and ways each interaction could happen, that every calculation using Feynman diagrams required 'heroic efforts of computation', says Jacob Bourjaily at the University of Copenhagen's Niels Bohr Institute.

This much became obvious in the 1980s, when the US was building the ill-fated Superconducting Super Collider in Texas. It was going to smash protons into each other, so it was imperative to understand the interaction of gluons, which hold together the quarks that make up protons. But it seemed impossible. 'Their complexity is such that they may not be evaluated in the foreseeable future,' one group of physicists wrote at the time.

Then there was an unexpected turn of events. In 1986, Stephen Parke and Tomasz Taylor from Fermilab near Batavia, Illinois, used Feynman diagrams and supercomputers to calculate the likelihoods of different outcomes for interactions involving a total of six gluons. A few months later, they made an educated guess at a one-line formula to calculate the same thing. It was spot on. More than 200 Feynman diagrams and many pages of algebra had been reduced to one equation, and the researchers had no idea why.

What was clear was that virtual particles were a big part of the problem. 'Every single Feynman diagram is a fantasy,' says Bourjaily. A fantasy in the sense that we have no way of observing the virtual particles they depict. What we do know is that the wild proliferation of mathematics required to account for them is very real, resulting in ridiculously unwieldy calculations.

Almost 20 years passed before another clue arrived. In 2005, Ruth Britto, Freddy Cachazo, Bo Feng and Edward Witten were able to calculate scattering amplitudes without recourse to a single virtual particle and derived the equation Parke and Taylor had intuited for that six-gluon interaction.

This time there was a lead on what the BCFW method might mean. It was inspired by a view of space–time called twistor theory, which had been developed in the late 1960s and early 1970s by Roger Penrose at the University of Oxford. The primary objects of this theory are not particles, but rays of light, or twistors. 'You can think of the universe as built up out of these rays, and points of space and time emerge at the places where these rays meet,' says Andrew Hodges, one of Penrose's colleagues at Oxford.

Hodges showed that the various terms used in the BCFW method could be interpreted as the volumes of tetrahedrons in twistor space, and that summing them up led to the volume of a polyhedron. The trouble was that his insight only worked for the simplest, most likely interaction of gluons with specific properties. For more complicated particle interactions, the resultant geometric objects were utterly bewildering. Their connection with real particle dynamics was intriguing, but the maths was too difficult.

It took Nima Arkani-Hamed and his team at the Institute of Advanced Studies (IAS) in New Jersey, including his then students Jaroslav Trnka and Bourjaily, to join the dots. Building on the seemingly esoteric work of pure mathematicians, the team arrived at a mind-boggling conclusion: the scattering amplitude calculated with the BCFW technique corresponds beautifully to the volume of a new mathematical object. They gave a name to this multi-dimensional concatenation of polyhedrons: the amplituhedron.

It's best to think of the amplituhedron not as a real object but as an abstraction. It's a mathematical structure that gives us an elegant way to encode the calculations that tell us how likely a particle interaction is to play out in a certain way.

The details of the interaction, meaning the number and properties of the particles involved and the forces involved, dictate the dimensions and facets of the corresponding amplituhedron – and that contains the answer. So there are actually many amplituhedra, one for each possible way in which a set of particles may interact.

The contrast with Feynman diagrams is stark. On one hand you may have to draw a thousand diagrams and use supercomputers, on the other you can get the same answer by calculating the volume of a single geometric object, even if the maths involved is far from trivial. 'It translates the physics problem into a purely mathematical problem – calculate the volume of that object,' says Trnka, who is now at the University of California, Davis.

It may transform physics, too – potentially nudging the door ajar to a unified theory of everything. That's because the amplituhedron does not embody unitarity and locality, those core principles baked into reality as described by Feynman diagrams. Scattering amplitudes that obey the laws of locality and unitarity do emerge from amplituhedra. But unlike in Feynman diagrams, the amplituhedron does not start with space–time that has these properties. 'The thing that you calculate will be unitary and local,' says Trnka. 'It's a consequence of the geometry.'

If so, locality is not a fundamental feature of space–time but an emergent one. That amounts to a radical rethink of reality, and one that could finally help us with a solution to one of the biggest questions in physics: how gravity behaves at the very smallest scales.

Locality and gravity don't sit well together. In order to precisely determine what happens at a given point in space–time, you have to zoom in closer and closer and examine smaller and smaller intervals of time. Quantum mechanics says that as one gets increasingly precise, the energy fluctuations in that region of space–time become bigger. Now, energy is mass, and mass has gravity, so incredibly high amounts of mass in a very tiny region of space ends up forming a black hole, which makes it impossible to see what's going on – and dashes any hopes of insight about the quantum nature of gravity. So, if gravity and quantum mechanics have to coexist, locality has to go.

The amplituhedron suggests that it can, potentially clearing the way for a quantum theory of gravity. That would finally let us understand what really goes on inside black holes and maybe even at the moment of the big bang – secrets of the universe that are theoretically impenetrable today.

If Arkani-Hamed is correct, that might just be the start. 'If we are going to lose something as dramatic as the idea of space–time, it's very unlikely that it leaves any of physics unaffected,' he told the audience at the String-Math 2016 conference in Paris. 'It must show up everywhere. It must show up even in situations where we think we understand things perfectly well.'

Naturally, there is a catch. Over the past few years, Arkani-Hamed and his colleagues have demonstrated that the amplituhedron works for a 'toy' model of particle interactions that involves supersymmetry, a theory in which all standard model particles have partner particles. But the standard model, our best description of reality, is not supersymmetric.

If that sounds like a killer blow, it isn't. 'The toy model is closer to reality than any other toy that people have played with over the last three decades,' said Arkani-Hamed in a talk at the IAS in 2017. Indeed, for some of the simplest, most likely particle interactions, the calculations using the amplituhedron agree with results obtained using standard calculation methods. Crucially, the new method holds for all four-dimensional theories of massless particles, supersymmetric or otherwise. The standard model has its origins in this class of theories, so it's entirely plausible that it will work there too. 'This correspondence with geometry is a general thing,' says Bourjaily. 'It's a statement about four-dimensional theories.'

Now the challenge is to extend this geometric way of thinking to more realistic models of particle interactions, and ultimately include gravity by doing away with locality. It's not going to be that simple, though. Which might be why Witten, who is also at the IAS, is simultaneously impressed and circumspect. 'Perhaps [the amplituhedron] is the closest we have to a unified picture, at least of some of the questions,' he says. 'There have been so many surprises in the study of these scattering amplitudes that it is rather hard to speculate on future directions. But it is pretty clear that there is a lot still to discover.'

Arkani-Hamed is confident that, ultimately, we will see that space–time and quantum mechanics emerge as one. 'In this baby example that's exactly what happens,' he said in Paris. 'There is no way in this geometry to decouple the piece which is space–time from the piece which is quantum mechanics. It's all one and the same aspect of the underlying positive geometry.'

## From $i$ to u: searching for the quantum master bit

*Our best theory of nature has imaginary numbers at its heart. Making quantum physics more real conjures up a monstrous entity pulling the universe's strings.* **Matthew Chalmers** *goes hunting for the grains of reality.*

If you've ever tried counting yourself to sleep, it's unlikely you did it using the square roots of sheep. The square root of a sheep is not something that seems to make much sense. You could, in theory, perform all sorts of arithmetical operations with them: add them, subtract them, multiply them. But it is hard to see why you would want to. All the odder, then, that this is exactly what physicists do to make sense of reality. Except not with sheep. Their basic numerical building block is a similarly nonsensical concept: the square root of minus 1.

This is not a 'real' number you can count and measure stuff with. You can't work out whether it's divisible by 2, or less than 10. Yet it is there, everywhere, in the mathematics of our most successful – and supremely bamboozling – theory of the world: quantum theory.

This is a problem, says respected theoretical physicist Bill Wootters of Williams College in Williamstown, Massachusetts – a problem that might be preventing us getting to grips with quantum theory's mysteries. And he has a solution, albeit one with a price. We can make quantum mechanics work with real numbers, but only if we propose the existence of an entity that makes even Wootters blanch: a universal 'bit' of information that interacts with everything else in reality, dictating its quantum behaviour.

What form this 'u-bit' might take physically, or where it resides, no one can yet tell. But if it exists, its meddling could not only bring a new understanding of quantum theory, but also explain why super-powerful quantum computers can never be made to work. It would be a truly revolutionary insight. Is it for real?

The square root of minus 1, also known as the imaginary unit, $i$, has been lurking in mathematics since the sixteenth century at least, when it popped up as geometers were solving equations such as those with an $x^2$ or $x^3$ term in them. Since then, the imaginary unit and its offspring, two-dimensional 'complex' numbers incorporating both real and imaginary elements, have wormed their way into many parts of mathematics, despite their lack of an obvious connection to the numbers we conventionally use to describe things around us. In geometry they appear in trigonometric equations, and in physics they provide a neat way to describe rotations and oscillations. Electrical engineers use them routinely in designing alternating-current circuits, and they are handy for describing light and sound waves, too.

But things suddenly got a lot more convoluted with the advent of quantum theory. 'Complex numbers had been used in physics before quantum mechanics, but always as a kind of algebraic trick to make the math easier,' says Benjamin Schumacher of Kenyon College in Gambier, Ohio.

## Quantum complications

Not so in quantum mechanics. This theory evolved a century ago from a hotchpotch of ideas about the subatomic world. Central to it is the idea that microscopic matter has characteristics of both a particle and a wave at the same time. This is the root of the theory's infamous assaults on our intuition. It's what allows, for example, a seemingly localised particle to be in two places at once.

And it turns out that two-dimensional complex numbers are exactly what you need to describe this fuzzy, smeared world. Within quantum theory, things like electrons and photons are represented by 'wave functions' that completely describe all the many possible states of a single particle. These multiple personalities are depicted by a series of complex numbers within the wave function that describe the probability that a particle has a particular property, such as a certain location or momentum. Whereas alternative real-number descriptions for something like a light wave in the classical world are readily available, purely real mathematics simply does not supply the tools required to paint the dual wave-particle picture.

## Hidden complexity

The odd thing is, though, we never see all that quantum complexity directly. The quantum weirdness locked up in the wave function 'collapses' into a single real number when you attempt to measure something: a particle is always found at a single location, for example, or moving with a certain speed. Mathematically, the first thing you do when comparing a quantum prediction with reality is an operation akin to squaring the wave function, allowing you to get rid of all the $i$'s and arrive at a real-number probability. If there is more than one way for a thing to end up with, say, a particular location, you add up all the complex number representations for each different way, and then square the sum.

The fact that this was a rather odd way to go about things first struck Wootters over 30 years ago, when he was writing his PhD thesis. In the 'real' world, the probability of rolling a 10 from a pair of dice is 3/36, because there are three ways to produce a 10 from the total of 36 possible outcomes. We add probabilities, we don't add them and square them. 'This procedure would be strange even if the square roots were real,' says Wootters.

The involvement of complex numbers only makes things worse. Complex numbers are two-dimensional with real and imaginary parts, whereas the probabilities of observable reality are only one-dimensional – purely real. That implies some of the information stored in the complex numbers disappears every time we make a quantum measurement.

This is very unlike the macroscopic, classical world, where there is at least in theory a perfect flow of information from the past to the future: given complete information about the properties and position of a pair of dice, for example, we can predict how they will fall.

Replace those complex square roots with real square roots, Wootters realised, and you can at least stem this troubling information haemorrhage. 'Real square roots would be comprehensible – not a strange thing – if nature is interested in a strong link between past and future,' he says.

As it turned out, he wasn't the first to try to make quantum theory real. In the 1960s, Swiss physicist Ernst Stueckelberg had reformulated quantum mechanics using only real numbers. But when he tried to impose the fundamental quantum principle of uncertainty – the idea that we cannot determine quantities such as the position and momentum of a particle simultaneously with full accuracy – he found that he still needed something to play the role of the imaginary unit.

Back in 1980, this was all a shrug-shoulders affair for Wootters. He had other fish to fry – namely, helping to found a new branch of physics, quantum information theory. This portrays quantum mechanics in terms of the information it can encode, and has allowed physicists to take advantage of fundamentally quantum processes such as teleportation and entanglement between particles to transport information more efficiently and securely than is possible by conventional, classical means. It also lies behind current attempts to build super-powerful quantum computers. Wootters's role as one of the founding fathers of the field has made him one of the most-cited physicists in the world.

It was an invitation in 2009 to give a talk in Vienna that prompted him to return to the problematic role of *i* in quantum theory. Could all the quantum information theory developed in the meantime provide any new resolution?

By October 2012, working with his students Antoniya Aleksandrova and Victoria Borish, Wootters had progressed far enough to think it could. Conventional quantum theory talks of information in terms of qubits – probabilistic versions of the traditional binary bits that classical computers crunch. Wootters and his colleagues were able to replace these qubits with real-number equivalents, and so capture all the weird correlations and uncertainties of conventional quantum theory without an *i* in sight. While he was about it, Wootters also published his 30-year-old insight that a real-number quantum theory could solve the problem of imperfect information flow during quantum measurements.

All this came with a huge sting in the tail, however. Like Stueckelberg's earlier attempt, this theory also needs an extra element to play the part of *i*. It turns out to be a monster: a physical entity Wootters dubs the u-bit.

A u-bit is a master bit: an entity that interacts with all the other bits describing stuff in the universe. Mathematically, this omnipresent conduit of information is represented by a vector on an ordinary, real two-dimensional plane. What it represents physically, no one, least of all Wootters, can tell, but by entangling itself with everything else in the universe, it is sufficient to replace every single complex number in quantum theory. The mathematical description of the u-bit supplies one further, slightly whimsical clue to its physical identity: whatever and wherever it is, it must be rotating quite fast.

It all sounds rather like a mistimed April fool, but that might just be the idea's strength, says Schumacher. 'I think a good paper in fundamental physics shares some characteristics with a good joke: it has an unexpected take on a familiar idea, and yet in retrospect it has a certain screwball inevitability. By that standard, Bill's u-bit theory is a very good joke.'

Matt Leifer of the Perimeter Institute in Waterloo, Ontario is more sceptical, pointing out that the nature of the u-bit makes it very difficult to test the theory. 'It's not yet clear whether it is something that can be manifested at a particular location such that you could find it in a detector, like we did the Higgs boson,' he says.

But the u-bit's influence might be felt indirectly. Since it interacts with everything in the universe, it can rough up even a supposedly isolated quantum system, making it 'decohere' and lose its vital quantum properties in a way not predicted by standard quantum theory. That could be bad news for the quantum computers built on the back of quantum information theory, which are very sensitive to environmental disturbance. With a u-bit lurking off-stage, physicists could try all they like to seal off such a device from the outside world, for example by putting it inside a huge fridge to cool down its interactions, but there would be nothing that could shield it from decoherence induced by the u-bit.

The length of time for which a conventional qubit could stave off decoherence is governed by the strength of the u-bit's interaction and its rate of rotation. An experiment carried out in 2011 by researchers at the National Institute of Standards and Technology in Boulder, Colorado, showed that qubits made out of trapped ions can remain quantum-coherent for a period of several seconds, a result that would suggest the u-bit interaction is weak, if it exists at all. 'The theory is definitely sensible, but it is currently not clear whether the u-bit is really realised in nature,' says Markus Müller of Heidelberg University in Germany.

For an information theorist like Wootters, it is easy to see the attraction of an entity that does away with the profligate waste of information seemingly hard-wired into the mathematics of quantum theory. But he is the first to admit that the u-bit is no panacea for quantum ills. It doesn't itself explain what wave-particle duality is, or why a particle can be in two places at once. Something like entanglement is actually more pervasive in the new formulation, because the u-bit can be fully entangled with any number of objects. In conventional, complex-number quantum mechanics, full-on entanglement is limited to two objects.

But even if the real-number reformulation just turns out to be a new angle from which to view quantum theory's mysteries, that could be valuable. Despite quantum theory's enormous success in agreeing with experiment, its conflict with our intuitions has left us casting round for a 'narrative' – a compelling explanation of why things should be the way quantum theory suggests they are. Often, those interpretations have been impeded by dwelling too much on one aspect of the theory, says Schumacher. 'There has been an amazing amount of nonsense written about quantum theory because people used to understand it only as a theory about waves, for instance.' That makes one insight from Wootters's work – that we should look at the information-carrying capacity of a quantum state to understand the physics behind it – an important one, says Müller.

As for the question of quantum theory's irreality, perhaps we have just to learn to love $i$. After all, it is not just quantum mechanics where its influence is felt. Complex numbers are also increasingly vital in describing optical waveguides, transitions between different states of matter and many other aspects of classical physics. 'People always thought of complex numbers just as a tool, but increasingly we are seeing that there is something more to them,' says mathematician Dorje

Brody of Brunel University in London. History tells us we have come to accept the reality of other mathematical concepts such as zero and the negative numbers only after long tussles. Perhaps 'imaginary' was an unfortunate choice of words when we came to name $i$. Unless, of course, instead of an $i$, a u-bit lurks, ubiquitous – and unseen.

## Reconstructing physics: the universe is information

*From virtual particles to imaginary numbers, reality never looked so shaky. And there's more. Leading quantum physicists **David Deutsch** and **Chiara Marletto** suggest abstract information is the key to understanding the universe – which places us at the centre of it.*

When we consider some of the most striking phenomena permitted by the laws of physics – from human reasoning to computer technologies and the replication of genes – we find that information plays a central role. But, on the face of it, information is profoundly different from the basic entities that physical sciences use to describe reality. Neither quantum mechanics nor general relativity, the most fundamental theories in physics, provide a meaning for information or even a way of measuring it. And it has a 'counterfactual' character: a message cannot carry information unless a different message is also possible.

Statements about information were therefore long regarded in physics as second-class, non-fundamental approximations. Information itself was considered an a priori abstraction, like Euclid's perfect triangles and circles, whose physical instantiations are inevitably approximate.

Yet there have long been clues that information is a

fundamental physical quantity, obeying exact laws. Consider statistical mechanics, pioneered by Ludwig Boltzmann at the end of the nineteenth century, which reformulates the laws of thermodynamics in information-like terms. For example, these laws define heat and entropy – loosely speaking, a measure of disorder in a system – in terms of the number of ways in which atoms of a given total energy could possibly be distributed, which is also the information content of the system. The laws of thermodynamics therefore link information with fundamental forms of energy, such as work.

Even more strikingly, in the 1970s Jacob Bekenstein and Stephen Hawking discovered that a black hole's surface area is also its entropy (in suitable units). Hence information, too, must be an exact quantity, like area.

In the theory of computation, information is referred to in the same way as the laws of thermodynamics refer to energy: without ever mentioning the details of the physical systems that instantiate it. Yet we now know that different laws of physics can give rise to fundamentally different modes of computing: quantum computers can solve problems qualitatively different from anything classical computers are capable of. Thus laws about computation must be laws of physics, and so must laws about information.

But what are those laws? How can abstractions be physical, and counterfactual properties factual?

We think we have solved this riddle. Our solution begins like this: the laws of physics have certain regularities that have never been expressed precisely, but only through a vague concept of 'information'. For instance, information is informally characterised as something that can be copied from one physical system to another – a property we call interoperability. A physical theory of information must express those regularities explicitly, in the form of fundamental laws. In this respect information is like energy, and its laws are like the principle of conservation of energy.

But unlike energy, the idea of information clashes with the prevailing conception of fundamental physics. Ever since Galileo and Newton, this has been that the physical world is explained in terms of its state (describing everything that is there) and deterministic laws of motion (describing how the state changes with time). Only one outcome can result from a given initial state, so there is no room for anything else to be possible. Information cannot be expressed that way, because of its counterfactual character. It requires a new mode of explanation, one provided by our constructor theory. Its basic claim is that all laws of physics can be expressed entirely in terms of statements of which tasks – i.e. physical transformations – are possible and which impossible, and why.

A task is possible if the laws of physics permit the existence of a *constructor* for it: something that can both cause the transformation and retain the ability to cause it again. A heat engine, for instance, is a thermodynamic constructor: it causes energy to change from one form to another, while operating in a cycle. A catalyst is a chemical constructor: it causes chemical reactions but is not itself chemically changed.

Just as is often done with catalysts, in constructor theory one abstracts away the constructor and expresses everything in the form of statements about tasks. Our constructor theory of information takes the informally known properties of information and expresses them entirely in terms of the distinction between possible and impossible tasks. This makes all the difference. In constructor theory, counterfactuals are first-class, fundamental statements, and transformations such as copying are naturally expressed as tasks. So in constructor theory the properties associated with information appear as elegant, exact laws of physics.

How is this achieved? First, one must express what it takes for a physical system to perform computations. All computations on a set of attributes of a system can be expressed as tasks – the permutations of that set. A *computation medium* is a system with a set of attributes whose permutations are all possible tasks. We call that a *computation set*. If copying the attributes in the computation set is also a possible task, we call the computation medium an *information medium*.

All the other laws about information can then be expressed in beautifully simple ways. For example, the interoperability of information is expressed as the principle that the combination of two information media is also an information medium.

Without any modification, this theory also expresses the properties of media capable of carrying out quantum computations. These properties turn out to define a species of information media that we call *superinformation media*. Given the counter-intuitive properties of quantum information (such as quantum cryptography, which is secure even against an eavesdropper who tries every possible key), one might guess that such media would be possible only by allowing some additional, weird class of tasks.

Remarkably, the opposite is the case. Superinformation media satisfy only a simple requirement: roughly speaking, that certain copying tasks on their states are impossible. This requirement gives rise to all the disparate features that distinguish quantum information from classical. So constructor theory explains the relationship between classical and quantum information, which has been poorly understood until now. It also reveals the single property underlying the most distinctive phenomena associated with quantum information, such as the unpredictability of where an electron lies despite the equations of motion being deterministic. This is a promising development in the quest to reveal exactly what is responsible for the power of a universal quantum computer.

Constructor theory has other far-reaching implications. One of the most fundamental is that the notion of knowledge can be expressed in objective terms, as information that can act as a constructor – such as, say, the program running on a computer that controls an automated car factory. In the prevailing conception, that idea is impossible to express, because one can only say what does or does not happen. In constructor theory, one must talk in terms of what is possible. And for almost any task that is possible under the laws of physics, the explanation of why it is possible is an account of how knowledge might be created and applied to build a constructor for that task.

That makes knowledge creators, such as people, central to fundamental physics for the first time since Copernicus debunked the geocentric model of the solar system.

## Does consciousness create reality?

*If placing humans at the centre of the universe is too much for you, look away now. **Douglas Heaven** says that while not a complete figment of our imagination, the universe may only become real because we're looking at it.*

Samuel Johnson thought the idea was so preposterous that kicking a rock was enough to silence discussion. 'I refute it thus,' he cried as his foot rebounded from reality. Had he known about quantum mechanics, he might have spared himself the stubbed toe.

Johnson was responding to Bishop Berkeley, a philosopher who argued that the world was a figment of our minds. Could he have been right? With its multiverses and cats both alive and dead, quantum mechanics is certainly weird. But some physicists have proposed that reality is even stranger: the universe only becomes real when we look at it.

This version of the anthropic principle – known as the participatory universe – was first put forward by John Archibald Wheeler, a heavyweight of twentieth-century physics. He likened what we call reality to an elaborate papier-mâché construction supported by a few iron posts. When we make a quantum measurement, we hammer one of those posts into the ground. Everything else is imagination and theory.

For Wheeler, however, making a quantum measurement not only gives us an objective fix on things but changes the course of the universe by forcing a single outcome from many possible ones. In the famous double-slit experiment, for example, light is observed to behave either as a particle or as a wave, depending on the set-up. The most baffling thing is that photons seem to 'know' how and when to switch. But this assumes that a photon has a physical form before we observe it. Wheeler asked: what if it doesn't? What if it takes one only at the moment we look?

Even the past may not yet be fixed. Wheeler proposed a cosmic version of the double-slit experiment, in which light from a quasar a billion light years away reaches us by passing around a galaxy that distorts its path, producing two images, one on either side of the galaxy. By pointing a telescope at each, observers would see photons travelling one of two routes as particles. But by arranging mirrors so that photons from both routes hit a detector at the same time, they would see light arrive as a wave. This time, the act of observation reaches across time to change the nature of the light leaving the quasar a billion years ago.

For Wheeler, this meant the universe couldn't really exist in any physical sense – even in the past – until we measure it. And what we do in the present affects what happened in the past – in principle, all the way back to the origins of the universe. If he is right, then to all intents and purposes the universe didn't exist until we and other conscious entities started observing it.

Sound crazy? Then try this one on for size. Another interpretation of quantum mechanics is Hugh Everett's many worlds hypothesis, which posits that everything that could happen has and does, in an infinite number of universes. Every time you make a decision, the universe splits in two, with you in one branch and an alternative you in the other, living the other possibility. The universe you occupy is, in some sense, an individual universe of your own making.

This idea is enough to give anyone a reality check. 'My natural inclination is to be a realist,' says Chris Timpson, a philosopher of physics at the University of Oxford. 'But if you're going to be a realist about the quantum world then you're left with a world that's very peculiar.' So peculiar, in fact, that the idea that it only exists because of us seems almost sensible.

# 5 It's a Mad, Mad, Mad, Mad Universe

For a long time, the skies must have seemed like an orderly place to humans: the sun rising and falling like clockwork, stars twinkling dutifully in their positions. The trouble began at least as far back as the Ancient Greeks, whose astronomers noticed some stars tracing out bizarre serpentine patterns in the night sky. They christened them *asteres planetai*, or wandering stars – now better known as the planets.

Since then, the view has only become more troubling. Earth was ousted from its position at the centre of the universe, and the universe itself was revealed to be a ballooning, billowing tapestry, embroidered with cosmological features that would have baffled the Ancient Greeks.

In this chapter, we visit some of the strangest places in the cosmos, where time flows backwards, white holes spew matter into existence, gravitational waves slosh around hidden dimensions and the universe itself leaks away through infinitesimal holes in the fabric of space–time. You'll never see the night sky in the same way again.

# Seven things you didn't know about the sun

*With its fiery rains, speedy magnetic flips and an atmosphere that defies the laws of physics, our home star is as weird as it gets, says* **Rebecca Boyle***.*

Billions and billions of stars fill our galaxy. Many burn bright, destined to become supernovae, while others are dim burnouts. They come alone and in pairs; with or without planetary companions. We have searched the far reaches of the universe in the hope of understanding the stars, but ultimately everything we know is based on our sole reference point, the sun. Yet our home star remains plenty mysterious.

'It's expected that it's understood, because it's right there, it's so close and dominant in the sky,' says astrophysicist Eamon Scullion from Trinity College, Dublin. 'How are we going to understand any other aspect of space if we can't get to grips with the nearest star?' While we may have to go back to square one, there are things we do know about our sun. It is made of plasma – gas that has been ionised, or highly charged. It fuses hydrogen in its core. It blasts us with radiation and, crucially, its life-giving light. As stars go, it is roughly middle-aged, having been around for 4.6 billion years. And it probably has 5 or so billion more to go before it swells into a red giant that consumes Mercury, Venus and Earth. Yet strange solar phenomena abound, and here are some of the strangest.

## It rains on the sun

We know the sun affects weather on Earth and in space, but it has its own dramatic weather phenomena, too. 'People have this image of a giant ball of gas that's on fire, and everything is streaming away from it at thousands of kilometres per second,' says Scullion. In fact, the sun's plasma can fall back to the surface as rain.

Though this so-called coronal rain was predicted about 40 years ago, we couldn't see or study it until our telescopes became powerful enough to spot it happening. It works a bit like the water cycle on Earth – where vapour warms, rises, forms clouds, cools enough to condense into a liquid and falls back to the ground as precipitation. The big difference is that the plasma doesn't change from gas to liquid, it simply cools enough to fall back down to the solar surface.

This all happens very quickly and on a gargantuan scale, with 'droplets' the size of countries plunging from heights of 63,000 kilometres – about one-sixth the distance from Earth to the moon. 'You basically generate something the size of Ireland in 10 minutes, and drop it out of the sky at a rate of 200,000 kilometres an hour,' Scullion says.

Solar tornadoes also form in a familiar fashion. Swirling solar plasma creates a vortex, which causes magnetic fields to twist and spiral around into a super-tornado that reaches from the surface into the upper atmosphere. Here they transfer energy and help to heat it, or so scientists believe.

## It has long-lost siblings

The sun may be on its lonesome now – its closest neighbour is 4.2 light years away – but that wasn't always the case. Once upon a time it had close family. After their birth in the same cloud of dust and gas that formed our solar system, these solar siblings scattered hundreds of light years apart in the Milky Way. In May 2014, astronomers reported the first one: a star called HD 162826.

'It looks like the sun, but a little bit bluer,' says Ivan Ramirez at the University of Texas at Austin, who led the study. It's also warmer than the sun and 15 per cent more massive. The star is about 110 light years away, and you can see it with the aid of a pair of binoculars in the left arm of the constellation Hercules.

To find its family ties, Ramirez's team combed through galactic archaeology studies, which model the motions of the Milky Way. These predictions laid out where sibling stars would be now if they had formed in the same place as the sun. Though they spread out in different directions, their positions still give away their birthplace, Ramirez says.

He narrowed down the search area to 30 stars, and then looked at them closely to find a family resemblance. Only HD 162826 had a similar chemical make-up to the sun. A separate team led by Eric Mamajek at the University of Rochester in New York also studied the star and found it is the same age as the sun, as would be expected for two stars born together. Even more tantalising, HD 162826 is already in a catalogue of stars that might harbour planets.

More than a dozen solar siblings have been identified since then. Locating them could tell astronomers more about the birth of our solar system, including what conditions were like when the sun and planets formed. But beyond scientific curiosity, Ramirez just wanted to find a member of the sun's nuclear family. 'It's a cool thing to do,' he says.

He plans to keep looking for more of our sun's lost litter-mates. Most are probably red dwarf stars, which are the most common stars in the galaxy. They are smaller and cooler than the sun, so they are much harder to find. But the *Gaia* tele-scope, launched in 2013, may help locate more solar siblings, as it will observe a billion stars to make the first 3D map of the Milky Way.

## It has a freaky calendar

Our planet's calendar is well known: it takes 24 hours to spin once on its axis – a day – and 365 days to travel around the sun – a year. Yet the sun's schedule is nothing like ours. Different parts of the sun spin at different rates. So while a day at the equator lasts 25 days, regions close to the poles take a few days longer to make a complete rotation. This uneven spin leads to distortion in the sun's magnetic field, which has knock-on effects. As the equator spins, it drags the magnetic field that connects the sun's poles, says Alex Young at NASA's Goddard Space Flight Center in Greenbelt, Maryland. This results in another strange calendar phenom-enon: solar maxima and minima.

As the sun's magnetic field gets wound up by the spin 'it starts to build tension and pressure, much like when you twist a rubber band and it knots up', Young says. Something has to give, so the magnetic fields snap and release energy in the form of heat, either as solar flares or furious clouds of energy called coronal mass ejections (CMEs).

This cycle, from magnetic twisting to energy releasing, happens over roughly 11 Earth years – giving the sun its own calendar. During what's called a solar minimum, flares are few and so are dark patches called sunspots that appear on the sun's surface due to intense magnetic fields.

## The sun's magnetic field reverses every eleven years or so

In solar maxima, more sunspots burst over the surface where they spew more flares and CMEs. Torrents of charged particles also stream through gaps in the sun's atmosphere and across the entire solar system. This can affect us, causing blackouts on Earth and damaging satellites. But each solar cycle varies, and we don't understand why, which makes them and their effects unpredictable.

The current cycle is unusually calm and has been one of the weakest since records began in 1755. This is in spite of some major solar storms, together with a colossal solar flare in 2012, which would have packed some punch had it hit Earth.

Predictions just a couple of years earlier suggested the cycle would be a scorcher, which shows just how little we understand solar cycles, says Todd Hoeksema, a solar physicist at Stanford University in California. 'It's like predicting the stock market. Past performance is no guarantee,' he says.

Also roughly every eleven years, the sun undergoes yet another calendar change: its magnetic field reverses. North becomes south, and vice versa. Earth does this, too, but only every 300,000 years or so (we are long overdue one). The sun's polarity last reversed in 2013, though the flip took scientists many months of analysis to confirm.

'Why is it 11 and 22 years and not 15 and 30? We don't know the answer to that yet,' Young says. 'When you think about it, it's such a short amount of time, given that the sun has been around for 4.6 billion years.'

## It breathes

As the sun follows its 11-year solar cycle, it changes, altering its output of solar wind, X-rays, ultraviolet and visible light. This has the knock-on effect of changing the size of the huge magnetic bubble of charged particles, called the heliosphere, that the sun blows around itself to way out beyond Pluto.

These changes affect everything from Earth's climate to the *Voyager 1* spacecraft, which finally entered interstellar space in 2013.

The sun provides nearly all the energy that drives Earth's climate – 2500 times as much as all other sources combined, according to Greg Kopp, a solar physicist at the University of Colorado's Laboratory for Atmospheric and Space Physics. In past epochs, solar cycles were partly responsible for warm periods and mini ice ages. Low solar activity drives cold winters in northern Europe and the US, and mild winters over southern Europe – although global warming means globally averaged temperatures are on the rise.

We now understand what's going on a little better thanks to a space-borne instrument called *TIM*, launched by NASA in 2003. *TIM* keeps tabs on the spectrum of energy the sun emits, and detects subtle changes in energy output so scientists can distinguish between human causes of climate change and purely natural causes we can't control.

Changes in the sun's output affect much more than just our climate, however. During a solar minimum, the solar wind streams from the poles at a much faster speed, so there's more pressure pushing against material from interstellar space. During solar maxima, the sun's magnetic fields are more knotted up and not as much wind escapes, so the heliosphere contracts. 'There's sort of an 11-year breathing,' says Hoeksema.

The solar wind has been 20 to 40 per cent weaker than expected this cycle, he says. This shallower breath is one reason why *Voyager 1* left the heliosphere earlier than scientists expected.

## It defies thermodynamics

Solar tornadoes are bizarre enough on their own, but they might help explain one of the sun's weirdest characteristics: its atmosphere is hotter than its surface. At 5700 kelvin, the sun's surface is scarcely cold, but it is frigid compared to the corona. The highest part of the sun's atmosphere, more than 1 million kilometres above the surface, can reach temperatures of several million kelvin.

Generally, an object cools as it moves away from a heat source; a marshmallow will toast faster when it's closer to a campfire flame than further away. But the sun's atmosphere does the opposite. Energy must be flowing into the corona, heating it up – but no one knows where this energy comes from. 'We don't fully understand the physics of what's going on,' Scullion says.

Computer visualisations might paint a clearer picture of this process – and quite artistically, too. In one simulation, NASA Goddard astrophysicist Nicholeen Viall added colour to data coming in from NASA's Solar Dynamics Observatory (SDO), which observed the sun's coronal plasma in 10 different wavelengths that each correspond to a temperature. The result is a swirling movie reminiscent of a Van Gogh painting. But Viall's visualisation suggested the atmospheric plasma was cooling, not heating. This may be because the heating is happening faster than SDO can detect.

Much of the energy that heats the corona appears to come from the so-called transition region – the area between the sun's corona and the next atmospheric layer down. Tornadoes, rain, magnetic braids, plasma jets and strange phenomena called 'spicules' are all thought to play a role in this heating process, bringing energy from the lower regions of the sun and depositing it higher up. But no one knows exactly how. NASA's Interface Region Imaging Spectrograph mission has been observing this region since 2013, and physicists like Scullion try to simulate these energy exchanges using models in the hope that they will yield clues that scientists can look for on the real thing.

## It's hard to get there

To truly understand all these solar conundrums, we need to get as close to the sun as possible. That's not as simple as flying straight there, as the operators of two new spacecraft that will fly closer to the sun than ever before are finding.

*Solar Orbiter* is a European Space Agency mission launching in 2018, aiming to fly within 45 million kilometres of the sun. It will photograph the sun's poles for the first time, which should help scientists understand how the sun generates its magnetic field, and may even give insights into why its magnetic polarity flips so frequently. By getting a close-up view, the probe will also be able to sniff the pristine solar wind, before it has reached Earth. The main goal is understanding how the sun interacts with the environment around it, says Tim Horbury, a physicist at Imperial College London and the principal investigator on the *Solar Orbiter*'s magnetometer. 'The basic physics is understood, but a lot of the detail is not,' he says.

NASA's *Parker Solar Probe* mission is set to launch shortly before ESA's mission and come even closer, just 6 million kilometres from the sun's surface. To get there, it will approach in a looping, circuitous route, like a matador approaching a wary bull. The slow approach is partly for safety's sake: as the probe gets closer, scientists can carefully monitor any threats from radiation or heat and adjust the approach if anything goes awry.

The *Parker Solar Probe* will lap around Venus seven times to put it on the right trajectory and also to build up speed and momentum to slingshot closer to the sun – at its closest approach, it'll zip past the sun at 200 kilometres per second.

Shielding a spacecraft from solar radiation is one of the most important jobs in space flight, but it's even harder when you are sidling up to the source. The technology to do it hasn't existed until now, Horbury says. Both craft will have beefy heat shields to protect their sensitive instruments from searing temperatures.

Both spacecraft will try to answer questions, including how the atmosphere is heated and how the sun generates its wind. But they will still be far from answering everything there is to know about our star, says Young. 'The problem is that you don't know what you don't know,' he says.

## White holes: hunting the other side of a black hole

*Black holes suck – but do they have mirror twins that blow? A far-flung space telescope is peering into galactic nuclei to spot one for the first time and may offer a gateway to parallel universes, writes* **Katia Moskvitch**.

Physics is full of opposites. For every action, there's a reaction; every positive charge has a negative; every magnetic north pole has a south pole. Matter's opposite number is antimatter. And for black holes, meet white holes.

Black holes are notorious objects that suck in everything around them. Famously, not even light can escape their awesome gravity. White holes, in contrast, blow out a constant stream of matter and light – so much so that nothing can enter them. So why have so few people heard of them?

One reason is that white holes are exotic creatures whose existence is speculated by theorists, but believed by few because no one has ever seen one. Now Nikolai Kardashev and his colleagues at the Astro Space Centre in Moscow are hoping to change that using a vast radio telescope with a view equivalent to that of a dish about 30 times wider than Earth. They are aiming to identify what lies at the heart of many galaxies. If they confirm the existence of white holes, they will cast into doubt our current notion of what lies at the centre of galaxies – including our own. It would also be vindication at last for physicist Igor Novikov, who was the first to theorise their existence in 1964. Back then, black holes were called frozen stars, and were even less well understood than they are today. Novikov did what theoretical physicists do when confronted with situations that are impossible to test in the laboratory: he used pure reasoning to ask what would happen to a black hole if time were to flow backwards. His thought experiment yielded a new kind of object that spewed matter and light continually: a white hole.

Others ran with the idea. What if a black hole was attached to a wormhole, a shortcut through space–time that connects two regions of our universe, or maybe even two different universes? The black hole would draw in matter, while at the other end of the wormhole there would be a white hole emitting it.

Many physicists, though, have found the notion of a white hole hard to swallow. After all, black holes are thought to form when a massive star collapses under its own gravity;

the collapsing matter results in a singularity at its core. This is the heart of a black hole, where all physical quantities diverge to infinity and all the known laws of physics break down.

But in the time-reversed version, 'a white hole existed in the past, and somehow exploded outward', says Novikov. Even he concedes the fundamental problem: 'Researchers accepted that, from a mathematical and theoretical standpoint, white holes could exist. But there were questions about how such an object could actually form.'

Wormholes offered a way, but there were theoretical problems with them, too. They seemed to collapse as soon as they formed, shutting down the white hole too. Novikov himself outlined this instability problem in the 1970s. A decade later, however, theoretical physicist Kip Thorne of the California Institute of Technology showed that wormholes could indeed be stable, which gave the white hole theory a boost. In 2014, Carlo Rovelli and Hal Haggard at Aix-Marseille University in France showed that quantum theory can transform a collapsing black hole into an expanding white hole.

Perhaps the fact that we have found no signs of a white hole, despite peering ever deeper into space, is a more fundamental problem. Enter a space telescope called *RadioAstron*, whose wildly elongated orbit takes it out to a distance of 350,000 kilometres – nearly as far as the moon and 30 times wider than Earth's diameter. Launched from Kazakhstan's Baikonur Cosmodrome in 2011, its dish is only 10 metres across. But when its signals are combined with those from radio telescopes on Earth, the resulting images are as sharp as those from a dish 350,000 kilometres wide.

Right now, *RadioAstron*'s resolution is 20 times better than that of any telescope on the ground. It is so good that it can pick out objects covering an angle of 27 microarcseconds – the size a snooker ball would appear on the moon as viewed from Earth. Kardashev and his colleagues have used *RadioAstron* to survey 100 active galactic nuclei, the compact regions at the centre of galaxies that are much brighter than expected. Many astronomers think that these owe their brilliance to supermassive black holes. As the black hole sucks in gas, the unlucky matter is sent swirling around and gets hot enough to sparkle before plunging into oblivion.

But could some of these dazzling displays instead be due to matter and light streaming out of a supermassive white hole? Novikov thinks so: 'Certain active galactic nuclei are not black holes, as most researchers suggest, but exist in the form of white holes, linking our universe to another universe.'

If *RadioAstron* can make a detailed enough image, then it should be easy to tell black holes and white holes apart, says Kardashev. 'If it's a black hole, then in the middle there should be a dark spot on the image,' he says. 'And if it's a white hole, then there should be a bright spot in the centre.'

But perhaps the reason we haven't seen a white hole is that we've been looking in the wrong place at the wrong time. Alon Retter, an astrophysicist who now works for Israel Aerospace Industries in Tel Aviv, thinks so. What's more, he believes that we may already have caught one flickering into existence.

In 2006, NASA's Swift satellite detected a gamma-ray burst called GRB 060614. Such bursts are usually associated with supernovae or regions of high star formation, but GRB 060614 was neither. Retter believes that it may instead have been a white hole. He argues that white holes will appear as temporary flashes, rather than shining continuously, because all that matter and light coming out will collapse under its own gravity into a black hole.

Kardashev and Novikov agree with Retter's ideas. 'The

nature of these flashes in the sky is still unclear,' says Kardashev. 'So once we spot a gigantic powerful gamma-ray burst with a lot of radio radiation, we will take a close look with *RadioAstron* and try to determine its shape and size for the first time.' That could provide important clues about its source. 'It may be a white hole or a wormhole. Maybe the flashes are coming from another universe.'

Retter calls his idea a 'small bang' – a spontaneous emergence of a white hole. If we extrapolate this thought, he says, we could assume that our entire universe is the result of a white hole that emerged as the big bang.

Hardly anyone is hunting for white holes these days, but hopefully that will change. With each passing day, *RadioAstron* is beaming back more observations of fine structures in active galactic nuclei. 'It is not theoretically excluded that the central engine in active galactic nuclei is something more interesting than a supermassive black hole,' says Konstantin Postnov, an astrophysicist at Moscow State University. 'So let's keep our eyes open and not discard even very exotic possibilities.'

## Five universal truths that might be wrong

*White holes aren't the only things breaking the laws of physics.* **Stuart Clark** *examines five cosmic impossibilities that just might turn out to be true.*

### The speed of light might not be constant

The speed of light in a vacuum is the ultimate cosmic speed limit. Just getting close to it causes problems: the weird distortions of Einstein's relativity kick in, so time slows down, lengths go up, masses balloon and everything you thought was fixed changes. Only things that have no mass in the first place can reach light speed – photons of light being the classic example. Absolutely nothing can exceed this cosmic max.

We have known about the special nature of light speed since an experiment by US physicists Albert Michelson and Edward Morley in the 1880s. They set two beams of light racing off, one parallel and one at right angles to the direction of Earth's rotation, assuming the different relative motions would mean the light beams would travel at different speeds – only to find the speed was always the same.

Light's constant, finite speed is a brake on our ambitions of interstellar colonisation. Our galaxy is 100,000 light years across, and it is more than four years' light travelling time even to Proxima Centauri, the closest star to the sun and home, possibly, to a habitable planet rather like Earth.

Then again, if the speed of light were infinite, massless particles and the information they carry would move from A to B instantaneously, cause would sit on top of effect and everything would happen at once. The universe would have no history and no future, and time as we understand it would disappear. We wouldn't like a universe like that.

But don't put the brakes on just yet. The fact is, a larger light speed would solve one of the biggest problems in cosmology: that the universe's temperature is more or less the same everywhere, even though there hasn't been enough time since the big bang for this thermal equalisation to have taken place.

Standard cosmology solves this problem with inflation, a period in the very early universe when space itself suddenly inflated faster than light speed (something Einstein's relativity does allow), carrying an equalised temperature to far-flung climes. But no one can find a plausible way for space to behave like this. Models of inflation have to be made flexible so they can retroactively fit just about any observation thrown at them.

You could achieve the same effect as inflation, however, if cosmic light speed started out infinite (or at least a lot larger) at the big bang and has been getting slower ever since as space has expanded. Initially, the speed fell precipitously. These days, it creeps downwards imperceptibly, explaining why we measure it as a constant.

That sounds wacky, but in 2016 Niayesh Afshordi at the University of Waterloo, Canada, and João Magueijo of Imperial College London proposed ways to test for a variable light speed in galaxy surveys or in fluctuations of the cosmic microwave background, the leftover radiation from the big bang. 'The idea that the speed of light could be variable was radical when first proposed, but with a numerical prediction, it becomes something physicists can actually test,' says Magueijo. 'If true, it would mean that the laws of nature were not always the same as they are today.'

And we should soon have answers. The HETDEX experiment at the recently upgraded Hobby–Eberly Telescope in Fort Davis, Texas, should soon start to provide data on the distribution of distant galaxies, as could the Dark Energy Spectroscopic Instrument experiment under construction at the Kitt Peak National Observatory in Arizona. Failing that, the next-generation CMB-S4 experiment should scrutinise the microwave background to the required accuracy. This alternative universe might not be too alternative at all.

## Quantum weirdness could be even weirder than we think

Imagine a world where, if you and I had once met, my missing the bus to work would automatically make you late too. Or where, if I put on odd socks, yours would be odd too. A great excuse, maybe – but also deeply weird.

The classical world we live in isn't like that. I do X and Y happens, and what Z is doing over there generally has little influence on that. But these clear relationships disappear when we enter the quantum world, the world of subatomic particles that are the building blocks of the universe – and encounter the phenomenon of entanglement.

Described by Einstein in 1935, this is a kind of particle telepathy that defies complete characterisation even today. Particles can become entangled when they interact, and once they do, no matter how far apart they are, measuring the properties of one automatically fixes the properties of the other – changes its socks, as it were.

Einstein decried this 'spooky action at a distance', yet many experiments have shown it is an essential ingredient of our world. 'Without quantum entanglement, we could not have quantum theory as we know it, and quantum theory is the basis of chemistry, our semiconductor industry, even life,' says Caslav Brukner of the Institute for Quantum Optics and Quantum Information in Vienna.

But here's the really weird thing. There's nothing stopping the quantum world having different levels of underlying correlation – largely uncorrelated worlds are possible within the broad sweep of the theory, as are ones that are far more connected. But only a universe with the exact level of weirdness that corresponds to entanglement produces the rich tapestry of phenomena, including life, that ours does.

So we probably shouldn't wish for any level of weirdness other than our own – but it would still be nice to know why things are as they are. Finding out how would probably mean deriving quantum theory from underlying principles like the constant speed of light, which is the foundation of Einstein's relativity. But the sheer universality of quantum theory makes this a far-off prospect, says Brukner. 'I'm not even sure that this goal can be achieved.'

According to quantum physicist Sandu Popescu of the University of Bristol, we may have to accept that such questions are not physical, but philosophical. 'We can predict exactly what will happen, but to say why it happens, we don't have a clue,' he says. 'It happens because nature is quantum mechanical – that is probably the best answer you will ever get.'

## The arrow of time could point both ways

If there's one thing that eats up time, it's working out what time is. It pops up in physical laws all over the place – but never quite as we expect it. In quantum theory, a 'master clock' ticks away somewhere in the universe, measuring out all processes. But in Einstein's relativity, time is distorted by motion and gravity, so clocks don't necessarily agree on how it is passing – meaning any master clock must, somewhat implausibly, be outside the universe.

Even odder, neither theory seems to place any restriction on time going backwards. The familiar one-way flow of time is expressed in only one area of physics: thermodynamics. If time flowed both ways, sometimes your coffee would warm up while sitting forgotten on your desk. Dropped eggs might spontaneously reassemble and leap from the floor into your hand. The dead might return to life and live backwards to birth, Benjamin Button style.

The culprit is entropy, essentially the thermodynamic measure of a system's disorder. When the universe was born, matter was randomly distributed throughout its tiny dimensions and it was the same temperature everywhere. Then gravity kicked in, pulling together matter and heating it up to form galaxies, stars, planets and other ordered imperfections.

Thermodynamics has been trying to re-establish disorder, increasing entropy every which way it can.

At a local level, entropy increase seems to be associated with information loss. Broken eggs do not reassemble because information about the former ordering is lost to us in the smash. You don't have all the information needed to put Humpty Dumpty together again – and that amounts to a barrier to travelling back in time.

Or does it? 'When the story is told like this it appears compelling, but the moment you start looking into more detail, it becomes more convoluted,' says Popescu. In classical physics, you could in principle reverse a thermodynamic process if you preserved the information by measuring the trajectories and velocities of all the components of a breaking egg – suggesting that we could reverse time.

So why can't we? One possibility, Popescu thinks, is an information gap intrinsic to the way the quantum world works. Here we are back with the phenomenon of entanglement. When a cup of coffee cools, Popescu believes, continual interactions between molecules of air and coffee increase their quantum entanglement. Although you can know what states an entangled particle pair contains, you can't definitively know which one has which state – leading to a continuous sapping of information from the world.

It's still just an idea, Popescu admits. 'Quantum mechanics is consistent with our macroscopic phenomenon being driven by quantum rules, but we cannot prove it,' he says. And there is a huge sting in the tail: if he's right, in some sense, time may be capable of flowing backwards after all.

That's because in a classical physics calculation, in theory all you need is a system's initial state and the laws of mechanics to work out what will happen for the rest of time. But in quantum mechanics, where a system's evolution is probabilistic, you can specify conditions for the initial state and final states of the system, and both of these conditions will influence the evolution. Apply this idea to the universe as a whole and 'information could be coming from plus infinity and propagating back through time', says Popescu.

There's no evidence of any of this so far, Popescu cheerfully admits. 'No one yet has investigated it seriously,' he says. 'It is speculative.' But if in the future physics shows that time really can travel backwards, well – in some sense we must already know.

## There may be extra dimensions all around us

We're accustomed to living in a three-dimensional universe. Well, four dimensions – time is a dimension too, albeit an oddly unidirectional one. But we've long thought there might be more large-scale spatial dimensions than the up-down, left-right, in-out we are all used to.

In the late nineteenth century, British mathematician Charles Howard Hinton suggested that what we perceive as different objects moving in relation to one another could be thought of as single, solid objects in a four-dimensional space passing through our three-dimensional universe. To get a sense of what that means, imagine what a spherical ball looks like observed as it passes through a two-dimensional sheet – as a circle whose radius expands and then contracts in time.

Adding extra dimensions to the universe is easy enough, on paper at least: you just need additional terms in your coordinate system. The question becomes how we perceive them. Einstein slipped in an additional space-like dimension to his equations of general relativity to explain how mass warps space–time. We don't perceive this dimension directly, but experience it as an acceleration and explain it as the force of gravity.

Some physicists are adamant that more physical dimensions must exist beyond those we can see. In string theory – still most physicists' chosen route to a unifying theory that combines gravity and the forces of the quantum world – the number of spatial dimensions is at least 10. This gives physicists enough wiggle room to try to explain all of the forces of nature together – but doesn't explain where these extra dimensions are.

Extra dimensions have some odd consequences, too – implying, for example, a multiverse of distinct universes next to one another. Not everyone likes that. 'I'm not a fan of the multiverse picture,' says physicist Erik Verlinde of the University of Amsterdam. 'Universes that we cannot communicate with are not that interesting to talk about. I think that we should be happy if we can explain the universe that we live in.'

Verlinde has been developing a quantum description of space and gravity to replace Einstein's smooth space–time 'continuum'. In his picture, minuscule building blocks made of quantum information become increasingly quantum entangled and create the seemingly continuous three dimensions of space.

But why three? That question remains open. 'A lot of these ideas can be implemented in two, three, four or higher dimensions, so I don't have an immediate reason why there should be three dimensions,' Verlinde says. And until someone can find one, tales of dimensions beyond those we can see might not be so wacky after all.

## There should be whole galaxies of antimatter out there

Antimatter has always been full of surprises. The first was that it existed. The second was that it didn't.

First things first. In the 1920s, British physicist Paul Dirac managed to marry quantum theory with Einstein's special relativity to explain how tiny, fast-moving fundamental particles such as electrons work. But his austerely beautiful unifying equation, honoured with a plaque in London's Westminster Abbey, had an unwanted consequence. For every matter particle like an electron, it predicted the existence of a second particle that was the same, but opposite in things like electric charge.

Dirac initially brushed this under the carpet – out of 'pure cowardice' he later said – but three years on, the antimatter version of the electron, the positron, was discovered in cosmic rays. Since then, as the standard model of particle physics was built on the foundation that Dirac and others laid, a very different problem has emerged. Antimatter shouldn't just exist, it should be abundant: every time a matter particle is made, an antimatter particle should also be conjured from the void. 'We should have a universe half full of antimatter,' says Michael Capell, an astroparticle physicist at the Massachusetts Institute of Technology. So where are these particles?

They can't be near us because matter and antimatter mutually 'annihilate' whenever they meet, and we would notice the flash of X-ray energy produced when they do. Various small-scale particle behaviours might allow there to be slightly more matter than antimatter, but none of these effects is nearly big enough to explain the size of the discrepancy we see.

Perhaps, then, the missing antimatter is elsewhere – in stars and galaxies made exclusively of the stuff, much as our sun and Milky Way are made solely of matter. Stars made of antimatter would give out the same light as ordinary stars, but also a wind of antiparticles, just as our sun gives out matter particles. When these antiparticles come into contact with ordinary matter outside their galaxy, they should produce X-rays that would again be visible across the universe.

We are yet to see anything of that ilk either, but the Alpha Magnetic Spectrometer (AMS) is performing a more direct test. This giant particle detector, lofted onto the International Space Station in 2011, can sort matter from antimatter in passing cosmic rays.

Positrons and antiprotons can be made relatively easily in today's universe, for example when high-energy particles collide in the strong magnetic fields around dead stars. The real prize would be something bigger. Most helium was made in the first few minutes of the universe's existence, so to find anti-helium could mean that the same process created the expected large quantity of antimatter. Stars are the only places where carbon and heavier nuclei can be made, so a single anti-carbon nucleus would confirm that there is an antimatter star somewhere in our universe.

It's like looking for a needle in a haystack – you would expect one complex antiparticle for every billion or so matter particles AMS detects, says Capell, who works on the project. The experiment has just about collected enough events to start saying something meaningful, but it is a race against time. The hunt has to be conducted in space, because antiparticles annihilate on contact with our atmosphere, but space is harsh on technology. 'AMS has been working like a champ but we can see that it is ageing,' says Capell. In 2014, one of its four cooling pumps stopped working – a worrying development for an experiment designed to last until 2024.

So fingers crossed for something soon to overturn the evidence of our eyes – that we live in an entirely matter-dominated universe.

## Antigravity: does anything fall upwards?

*If a substance could resist gravity, it would rewrite physics textbooks. Amazingly fiddly experiments to test whether antimatter can are kicking off.* **Joshua Howgego** *meets the physicists turning the cosmos upside down.*

On 11 November 2016, a small birthday party was held in an apparently unremarkable hangar on the outskirts of Geneva. Nothing too fancy, just a few people gathered around a cake. The honourees were there. Well, sort of – they were still locked in the cage where they had spent their first year. But then again, there is no other way to treat a brood of antimatter particles.

The antimatter realm is so bizarre as to be almost unbelievable: a mirror world of particles that destroy themselves and normal matter whenever the two come into contact. But it's real enough. Cosmic rays containing antiparticles constantly bombard Earth. A banana blurts out an anti-electron every hour or so. Thunderstorms produce beams of the stuff above the planet.

Making and manipulating antimatter ourselves is a different kettle of fish. Hence that birthday party held at the particle physics centre CERN, celebrating on behalf of a quadruplet of antiprotons. There's a lot we would like to learn from these caged beasts and their ilk, not least this: do they fall up?

Cards on the table, few physicists believe that such 'antigravity' effects exist – that if you released one of those antiprotons and somehow ensured it free passage through the hostile world of matter, it would magically float up. But the recalcitrant nature of antimatter means we've never done the experiments, and until we do, we simply don't know. 'Progress is often made by asking the questions we think we already know the answer to,' says Daniel Kaplan of the Illinois Institute of Technology in Chicago.

The scepticism about all forms of antigravity dates back to the 1950s, when the physicist Hermann Bondi was pondering the implications of general relativity, Einstein's theory of how gravity arises from warping the fabric of the universe. Gravity is an odd sort of force, not least because it only ever works one way. With electromagnetism, say, there are positive and negative charges that attract and repel. With gravity, however, there are only positive masses that always attract.

Bondi showed what a bizarre world it would be if this were not the case, demonstrating how negative mass would end up pursuing positive mass across the universe. This sort of 'runaway motion' does not appear to exist – but we should be careful about what we draw from that, says Sabine Hossenfelder of the Frankfurt Institute for Advanced Studies. 'People who speak of the runaway problem often jump to conclusions from Bondi's argument and conclude that anti-gravitation itself is inconsistent,' she says. 'But it merely requires a modification of general relativity.'

And here's the thing: general relativity is probably due a modification. The theory is incompatible with quantum mechanics, the other great pillar of modern physics, and if we are to find a way to make a unified description of the universe, that must change. Then everything is up for grabs.

So in a few labs around the world, the search for negative mass and its associated effects goes on. Antimatter is a particularly promising place to look. It is just like normal matter but with the opposite electric charge and a few other mirrored quantum properties. There's no reason to think it has the opposite mass and anti-gravitates, and some good reasons to think it can't have.

But if antimatter did anti-gravitate, that might help with another of its central mysteries: where most of it is. Our theories say matter and antimatter should have been created in equal proportions in the big bang, and yet we live in a matter-dominated world.

## The emptiest box

Explaining this glaring inconsistency has largely been a case of trying to find asymmetries in the processes of particle physics that favour normal matter. Such asymmetries do exist – but they are about a trillionth of the size needed to explain matter's supremacy. 'People have been trying to make it work – and it doesn't work,' says Kaplan.

Antigravity could provide a better explanation. A repulsive gravitational interaction could have driven matter and anti-matter away from each other so they never had the chance to annihilate in the early universe. Since then, the ongoing expansion of the universe would have driven the twain ever farther apart – and the antimatter might eventually have created its own galaxies in other corners of the universe. 'Then the missing antimatter would be hiding in plain sight,' says Kaplan's colleague Thomas Phillips.

Add to that the technological possibilities that levitating matter away from Earth's surface might bring, and even the US air force wants in – it has given millions of dollars to antimatter researchers over the years. Unfortunately, doing the experiments turns out to be quite an ask.

The problems start with needing a home for antimatter that is almost entirely free of normal matter. That requires some of the emptiest boxes on Earth, containing just hundreds of gas molecules per litre (there are about $10^{22}$ in a typical litre of air). But even these boxes have sides. To stop the antimatter banging into them and instantly annihilating, you must slow it down by cooling it to within a few degrees of absolute zero and then catch it in a vortex of electromagnetic fields. Little by little, we've been perfecting these arts, holding antimatter particles for seconds, minutes, days – and for a year, as celebrated at the November 2016 party.

That milestone was reached by CERN's Baryon Antibaryon Asymmetry Experiment (BASE), one of six experiments competing to measure antimatter's fundamental properties that are all housed in CERN's vast Antimatter Deceleration Hall. Inside, past a sign marked 'Antimatter factory', the most noticeable things are the bright yellow cranes, swinging around the vats of liquid nitrogen required for cooling. Somewhere below, a beam of particles from CERN's Proton Synchrotron accelerator smashes into a block of metal, creating a plethora of particles. A system of magnets selects the antiprotons and funnels them into a ring of more magnets that keep them on course as they are decelerated for trapping.

Experiments have been running here since the 1990s, studying whether antimatter and matter particles truly are as close to identical as we think. In 2015, by measuring how antiprotons danced around in a magnetic enclosure known as a Penning trap, BASE produced the most precise measurement yet of their mass-to-charge ratio. They showed it was the same as a proton's, to about 69 parts per trillion, four times more precise than the previous best value. In late 2016, the neighbouring ASACUSA experiment produced the most accurate measurement yet of the antiproton's mass, finding no evidence of a different value from the proton's.

The same value – but is the mass positive or negative? That is the multimillion dollar question, and it takes the experiments to a new level of fiddliness. Gravity is weak and easily overwhelmed by the electromagnetic force, so using charged particles such as antiprotons and controlling them with magnetic fields won't do. You could try getting an antiproton in position and shutting off the magnets to see which way it falls, but the antimatter's electrostatic interactions with its surroundings would overwhelm any gravitational push or pull it might feel.

A better bet is neutral atoms of antimatter, such as anti-hydrogen. Making these is no cakewalk, but they have a tiny electric polarity that makes it worth going the distance – their electrostatic interaction isn't strong enough to swamp gravity, but very strong magnetic fields will still hold them in place. CERN's Antihydrogen Laser Physics Apparatus (ΛLPHA) experi-ment has been doing this since 2005, and now routinely traps and holds bunches of antihydrogen atoms for about 15 minutes. 'Just the other day we trapped 350,' says Jeff Hangst, head of ALPHA.

## Not up, just less down

In 2013, ALPHA published a proof of principle measurement, briefly collecting a cloud of 434 antiatoms, turning off the magnets and tracking their subsequent motion by where they annihilated. It was a crude test, and inconclusive – the final answer was compatible with the antiparticles having either negative or positive gravitational mass.

Work on a souped-up version that gives the particles more space to fall started in 2018. 'We're going to knock out a wall and build a vertical version of the experiment next door,' says Hangst. Getting the necessary accuracy won't be easy, because the antiatoms ALPHA uses are relatively hot and so jiggle around, which clouds the issue. But large enough numbers of antiatoms should help us answer the central question. 'Up or down – that should be possible,' says Hangst.

A further CERN experiment, AEGIS, also aims to perform

tests within a few years. Kaplan is planning experiments with muons, heavier cousins of the electron, and a team led by David Cassidy of University College London is planning to use positronium, an 'atom' consisting of an electron and its antimatter partner, a positron, orbiting one another.

Back at CERN, the Gravitational Behaviour of Antimatter at Rest, or GBAR, experiment intends to tackle the question using a single antihydrogen ion, a combination of one antiproton and two positrons. In theory, it should be easy to hold this charged speck in place with magnetic fields and cool it with lasers. The idea is then to knock off a positron using another laser, making the antiatom neutral. At this point it would cease to feel the effect of the trapping field and fall – up or down. GBAR's head, Patrice Perez, says they expect to make measurements sensitive to detect even a 1 per cent deviation from the gravity felt by normal matter.

Construction of the experiment requires new lasers and an extra antiproton decelerator called ELENA. Hangst is confident of beating the upstart to the punch. 'I view GBAR as a case of five miracles happen and then it works,' he says. One telling fact is that GBAR plans on using only one detector, below the trap. 'We really do not expect antimatter to fall up,' says Perez.

Even if it falls at all differently, however, that would still be hugely interesting. 'In all the descriptions I know, antimatter cannot antigravitate,' says Sergey Sibiryakov of CERN. What's more plausible, he thinks, is that there might be other forces that modify gravity whose effects cancel out on normal matter, but not on antimatter. In that case, antimatter might not fall up – just less down. 'Now, that's not natural, but it is logically possible,' he says. Similar gravity-modifying effects might be produced if the graviton, a quantum particle proposed to carry the force of gravity, has a small mass, rather than being massless as is usually assumed.

Even so, we probably shouldn't be holding our breath for amazing self-levitating machines any time soon. A more immediately practicable way of using antimatter to beat gravity might be to harness the energy released when it annihilates. One firm, Positron Dynamics in Livermore, California, has been developing the idea with financial support from PayPal co-founder Peter Thiel, among others.

Positron-fuelled rockets could power spacecraft much further and faster than is currently possible, according to Positron Dynamics co-founder Ryan Weed. 'Our vision is to create technology that allows humanity to venture outside of our solar system,' he says. The company's patented system involves harvesting positrons from radioactive sodium-22 and using these to start off a nuclear fusion reaction that generates thrust. Weed says the team is set to test the device in a lab and wants to test it in orbit in the next few years.

But experience makes Stefan Ulmer, the head of the BASE experiment, cheerfully sceptical of immediate progress. Antimatter won't be easily tamed. 'In the whole history of the CERN Antimatter Deceleration Hall, we've produced about enough to heat up a cup of water by about 5 degrees,' he says. Not even enough, in other words, to make a pot of tea to wash down that birthday cake.

## A quantum leak could be flooding the universe with dark energy

*The loss of countless tiny drops of energy since the start of the universe might be behind the rising tide of dark energy accelerating the cosmos's expansion, writes **Joshua Sokol**.*

If physicists went in for commandments, the first would surely be: thou shalt not get something from nothing. Also known as the principle of energy conservation, this universal accounting law makes it impossible for energy to be magicked either into or out of existence.

Whenever a suspicious transaction seems to take place in physics, a careful audit with the principle of energy conservation usually reveals the source of the error – some overlooked entry in the ledger that, once taken into account, helps balance the books.

This time-honoured technique has allowed us to predict planets and discover particles. But now it appears to be under attack. Look out into the depths of the universe today, and you see a vast quantity of energy. It is so vast, in fact, that it accounts for over two-thirds of all the energy there is. And this mysterious stash is growing continuously – energy laundering on the grandest of cosmic scales.

Working out where this so-called dark energy comes from is probably the biggest problem in physics. We have long been frustrated in finding a solution, but now two groups of physicists think they have it. If they are right, we may have found dark energy's source in the imperfect joins of the universe where different theories of reality meet. Follow the trail back, and we could even arrive at a better theory of reality.

We've known about the invisible elephant in the universe for some time. In 1998, astronomers observing distant supernovae noticed that they were even dimmer than expected. We expected their light to fade as it travelled towards us across an expanding universe, but these new results suggested there was a foot on the accelerator.

Dark energy is the mysterious substance conjured up to explain what is pushing the universe apart ever faster. And as the latest data from sources like the Planck space telescope reveal, it is spread evenly throughout the universe at a density equivalent to around half a dozen protons in every cubic metre of space.

The simplest way to explain this all-pervasive energy is to think of empty space as not being empty after all. On very small scales, quantum mechanics says that any vacuum is filled with the wriggling of quantum fields. But calculations following that approach give us a dark energy density that is 120 orders of magnitude larger than the one astronomers measure from the accelerating expansion of the universe. Almost laughably wrong.

Some researchers, however, haven't given up on making these two numbers square. According to a new paper by Qingdi Wang, a student of theoretical physicist Bill Unruh at the University of British Columbia, these jiggling fields would tend to cancel each other out on larger scales, drastically deflating the prediction.

But frustration at the inability to make progress has now led some to suggest that it's all down to a cosmic accounting error. The idea is that dark energy is not actually a substance held in the universe's vaults – it's something that appears on the books purely because there's something else we've overlooked.

Energy conservation is such a basic principle that any apparent violation should give pause for thought. The seminal work of mathematician Emmy Noether in the early twentieth century showed that energy conservation was an expression of something even more fundamental: the idea that the laws of physics are immutable over time. And indeed, it is a principle that has paved the way for centuries of discovery.

When it comes to dark energy, cosmologists already had a vague idea where the accounting error might lie. According to Einstein's equations of general relativity, energy is absorbed and released all the time by the bending and stretching of the fabric of space–time. When photons seem to lose energy as they travel across an expanding universe, for example, that energy is all assumed to go into the universe's geometry. On the scale of the cosmos as a whole, energy is always appearing to be either created or destroyed.

Similarly, dark energy isn't adding or subtracting anything from the universe's overall budget. From afar, cosmologists are confident that everything balances out between the universe's stuff and the warped space–time holding it. But up close, the exact nature of the transaction bestowing space with extra energy remains mysterious. 'The question is "Where is it coming from?"' says Spiros Michalakis at the California Institute of Technology.

## Secret source

Such a grainy structure would have repercussions for the objects that inhabit it. Relativity dictates that particles with mass bloat or compress the space around them depending on how much mass they have. The process is often equated to a taut sheet bending under the influence of a bowling ball rolling around on top of it. But what if up close, the sheet's surface was stippled?

In such a situation, Josset and his colleagues argue, particles are likely to feel that graininess as a form of friction, shedding energy into the stitching of space. If their model holds, the matter in the universe has been losing energy continuously since a fraction of a second after the big bang.

Adding up the little losses of energy between then and now gives an estimate for dark energy's strength closer to reality than the 120 orders of magnitude overestimate, although still quite a way off. 'We are only seven orders of magnitude away,' says Josset's colleague Alejandro Perez of

Aix-Marseille University, noting that they plan to keep refining their estimate.

For Thibaut Josset of Aix-Marseille University, the processes responsible for that transaction lie in the jagged edges where quantum mechanics and general relativity meet. For decades, we have been looking for a unified theory of quantum gravity, one capable of explaining microscopic quantum processes alongside the large-scale workings of gravity. Thus far, no such theory exists.

One key difference between general relativity and quantum mechanics lies in the way they see the universe's fundamental structure. In Einstein's view, which works perfectly for objects on the scale of planets, stars and galaxies, the four dimensions of space and time are smooth and continuous. But quantum mechanics, which seems to govern reality at small scales, implies that deep down, space, like everything else, must be made up of discrete units that we still don't know how to describe.

'The magic of this thing is that very tiny violations of energy conservation, that are very, very hard to detect in normal, local experiments, build up during the very long history of the universe,' says Perez. Add them all up, and you could have enough to explain away dark energy. In other words, it is the tiniest drip-drip of energy – the smallest of leaks in space–time – that is causing this biggest of problems to accumulate.

The leak would have to be so small as to have gone unnoticed so far. At the Large Hadron Collider and elsewhere, experimental physicists are on the lookout for apparent violations of conservation of energy, as spotting one might indicate the existence of new particles. So far they haven't found any good leads, and the chances are they won't with current particle colliders. But the amount of energy non-conservation these experiments allow is still enough to hide the observed strength of dark energy, Perez says, something like the mass of a proton going missing every year from a cube of water 10 kilometres across.

While physicists have long known about general relativity's ability to transfer energy in and out of the space–time curvature on a grand scale, they have struggled to make it work on the scales Josset describes. What has stood in their way, says Sabine Hossenfelder, a theoretical physicist at the Frankfurt Institute for Advanced Studies, is that Einstein's equations are ruthless about energy conservation when you zoom in on small regions of space. Any quantum jiggery-pokery would invalidate the mathematics. That is, until Josset's colleagues suggested using a less restricted view that Einstein himself had worked on. This workaround allowed Josset to relax restrictions on energy conservation. 'I'm annoyed I didn't think of it earlier,' says Hossenfelder.

But one of Josset's assumptions remains contentious. The idea that space–time is ultimately made up of grains, while popular, is far from proven. Identifying the source of dark energy in the interplay between quantum theory and general relativity may require a different approach. Natacha Altamirano of the Perimeter Institute in Waterloo, Canada, and her colleagues have come at the problem from a different angle. Or rather from the largest possible scale, to examine how quantum mechanics and general relativity play off each other across the entirety of the universe.

Altamirano's work considers what would happen to a particle traversing the smooth hills and valleys described by Einstein's theory, but within a universe itself following the fuzzy rules of quantum mechanics. Considering a quantum universe is a familiar gambit in theories of quantum cosmology, which try to explain the universe's earliest instants back when it was still tiny and ruled by wild fluctuations. If the whole universe was quantum, then, much like an electron orbiting an atom, the cosmos could theoretically exist as a superposition of many different possible sizes and states at once.

In practice, the universe's choices are a lot more limited. The reason lies in Heisenberg's uncertainty principle, which governs the precision with which we can know the value of any quantum variable. Measure the position of a particle very accurately, for example, and you can't closely measure its momentum, and vice versa. As photons travel from one galaxy to another, they lose energy. And in the language of the uncertainty principle, that's a lot like taking a cosmological measurement.

All those unintentional measurements of the universe force the quantum uncertainty to go somewhere else. And one of the ways that can manifest itself is in the form of information loss elsewhere: a little more uncertainty in the rate the universe is accelerating, for example. That change, in turn, would have consequences for all other variables that depend on that rate, further changing the acceleration of the universe in a perpetual feedback loop.

Unless you accounted for it, all that noise would add up to a mysterious dark energy-like term popping out of the void. 'If I decide to describe my universe with a theory of general relativity that conserves energy, I would see this extra fluid,' says Altamirano. As to whether the dark energy density that emerges from such a model might turn out to match reality in a way that rivals Josset's, Altamirano is still working on generating such a figure. 'I can't tell how possible that is,' she says.

Josset's and Altamirano's approaches come at a time when dark energy has repeatedly foiled theorists' attempts to nail it down. Not everyone is convinced the pair are barking up the right tree, though. For Antonio Padilla at the University of Nottingham, the sheer difference in scales makes it unlikely that quantum gravity effects on the smallest imaginable sizes can explain dark energy, which is manifest across billions of light years.

They can't both be right, either. Because the two approaches use different mathematical language to discuss how gravity and the quantum world interact, they produce different answers for what happens to cosmology. Altamirano's model produces something like dark energy, but it's a kinder, gentler version. As the universe expands, it dilutes in space, whereas the dark energy density predicted by Josset's model remains a constant, in keeping with observations.

Ironing out such wrinkles to everyone's satisfaction would probably need a fully formed theory of quantum gravity – or at least some as-yet unimagined experimental test that would allow us to look at the universe's very earliest moments. Until then, dark energy will continue to accumulate interest in the distant reaches of the cosmos – a silent rebuke to the idea that we have our cosmic accounting practices in hand.

## Gravitational waves could reveal hidden dimensions

*Signatures of extra dimensions that don't normally affect the four dimensions we can observe could show up in the way they warp ripples in space–time, writes* **Leah Crane**.

Hidden dimensions could cause ripples through reality by modifying gravitational waves – and spotting such signatures of extra dimensions could help solve some of the biggest mysteries of the universe.

Physicists have long wondered why gravity is so weak compared with the other fundamental forces. This may be because some of it is leaking away into extra dimensions beyond the three spatial dimensions we experience.

Some theories that seek to explain how gravity and quantum effects mesh together, including string theory, require extra dimensions, often with gravity propagating through them. Finding evidence of such exotic dimensions could therefore help to characterise gravity, or find a way to unite gravity and quantum mechanics – it could also hint at an explanation for why the universe's expansion is accelerating. But detecting extra dimensions is a challenge. Any that exist would have to be very small in order to avoid obvious effects on our everyday lives. Hopes were high (and still are) that they would show up at the Large Hadron Collider, but it has yet to see any sign of physics beyond our four dimensions.

In the last few years, though, a new hope has emerged. Gravitational waves, ripples in space–time caused by the motion of massive objects, were detected for the first time in 2015. Since gravity is likely to occupy all the dimensions that exist, its waves are an especially promising way to detect any dimensions beyond the ones we know.

'If there are extra dimensions in the universe, then gravitational waves can walk along any dimension, even the extra dimensions,' says Gustavo Lucena Gómez at the Max Planck Institute for Gravitational Physics in Potsdam, Germany. Lucena Gómez and his colleague David Andriot set out to calculate how potential extra dimensions would affect the gravitational waves that we are able to observe. They found two peculiar effects: extra waves at high frequencies, and a modification of how gravitational waves stretch space.

As gravitational waves propagate through a tiny extra dimension, the team found, they should generate a 'tower' of extra gravitational waves with high frequencies following a regular distribution. But current observatories cannot detect frequencies that high, and most of the planned observatories also focus on lower frequencies. So while these extra waves may be everywhere, they will be hard to spot.

The second effect of extra dimensions might be more detectable, since it modifies the 'normal' gravitational waves that we observe rather than adding an extra signal. 'If extra dimensions are in our universe, this would stretch or shrink space–time in a different way that standard gravitational waves would never do,' says Lucena Gómez.

As gravitational waves ripple through the universe, they stretch and squish space in a very specific way. It's like pulling on a rubber band: the ellipse formed by the band gets longer in one direction and shorter in the other, and then goes back to its original shape when you release it.

But extra dimensions add another way for gravitational waves to make space shape-shift, called a breathing mode. Like your lungs as you breathe, space expands and contracts as gravitational waves pass through, in addition to stretching and squishing. 'With more detectors we will be able to see whether this breathing mode is happening,' says Lucena Gómez.

'Extra dimensions have been discussed for a long time from different points of view,' says Emilian Dudas at the École Polytechnique in France. 'Gravitational waves could be a new twist on looking for extra dimensions.' But there is a trade-off: while detecting a tower of high-frequency gravitational waves would point fairly conclusively to extra dimensions, a breathing mode could be explained by a number of other non-standard theories of gravity. 'It's probably not a unique signature,' says Dudas. 'But it would be a very exciting thing.'

# 6 At the Limits of Imagination

Quick, what's the biggest number you can think of? Infinity? Infinity plus one? Mathematicians struggle with this problem all the time; luckily they've found ways to do calculations on numbers that are too big to ever write down.

They're not the only ones having to stretch their minds around some unfathomably big numbers. In this chapter, we'll explore some of the most mind-boggling extremes around. Our journey begins on a scale that has no end, climbs the millimetre-high mountains on the surface of neutron stars and samples a handful of atoms so hot they exist at negative degrees kelvin. Get ready for some seriously sweaty mental gymnastics – it's Bikram yoga for the brain.

# At the Limits of Imagination

# The biggest number imaginable

*Infinity is larger than large. It's only when you try to work out how much larger that you realise some infinities are larger than others.* **Richard Webb** *starts counting.*

Ian Stewart has an easy, if not particularly helpful, way of envisaging infinity. 'I generally think of it as: (a) very big, but (b) bigger than that,' says the mathematician from the University of Warwick. 'When something is infinite, there is always some spare room around to put things in.'

Infinity is one of those things with a preprogrammed boggle factor. Mathematically, it started off as a way of expressing the fact that some things, like counting, have no obvious end. Count to 146 and there's 147; count to a trillion and say hello to a trillion and one. There are two ways of dealing with this, says Stewart. 'You can sum it up boldly as "there are infinitely many numbers". But if you want to be more cautious, you just say "there is no largest number".'

Only in the late nineteenth century did mathematicians plump for the first option, and begin to handle infinity as an object with properties all its own. The key was set theory, a new way of thinking of numbers as bundles of things. The set of all whole numbers, for example, is a well-defined and unique object, and it has a size: infinity.

The sting in the tail, as the German mathematician Georg Cantor showed, is that by this definition there is more than one infinity. The set of the whole numbers defines one low-lying sort, known as countable infinity. But add in all the numbers in between, with as many decimal places as you please, and you get a smoother, more continuous infinity – one defined by a set that is infinitely bigger.

That is just the beginning. Hugh Woodin is a set theorist at Harvard University who has a whole level of infinity named after him, a particularly vertiginous level populated with numbers known as Woodin cardinals. 'They are so large you can't deduce their existence,' he says.

Such infinities help solve otherwise unsolvable problems in less rarefied mathematical landscapes below. They are the ultimate abstraction: although you can manipulate them logically, you can't write formulae incorporating them or devise computer programs to test predictions about them. Woodin's notepads consist mainly of cryptic marks he uses to focus his attention, to the occasional consternation of fellow plane passengers. 'If they don't try to change seats they ask me if I'm an artist,' he says.

How closely our common-sense conception of endlessness matches the mathematical infinities isn't clear. But if we can't quite grasp boundarylessness, it probably doesn't matter, says Woodin – however you slice it, infinity seems far removed from anything we see in the real world. Perhaps those enigmatic markings aren't so different from those of his fellow passengers after all. 'It might be we're just playing a game,' says Woodin. 'Perhaps we are just doing some glorified sudoku puzzle.'

# The most extreme places in the universe

*Now that you've wrapped your head around infinity, you should have no problem taking in the improbably huge numbers needed to measure these seven cosmic record breakers.* **Stephen Battersby** heads out on an interstellar safari.

## The hottest thing in the universe

A journey towards the hottest climes of the cosmos must start by passing the sun, the fiery centre of our solar system. With a surface temperature of 5700 kelvin, our star is far from chilly, but it is no cosmic record breaker either. Blue supergiants, whose greater mass compresses their cores and stokes the nuclear fires within, run at more than 50,000 K.

Even that is surpassed by some white dwarfs, compact spheres of heat left behind when a smallish star burns out. One such stellar cinder, called HD62166, measures a scorching 200,000 K and lights up a vast nebula with its painfully bright atmosphere.

Plunging deep inside a star will take you to even more hellish realms. The largest supergiant stars may have core temperatures of more than a billion kelvin. For a stable star, the theoretical upper limit is about 6 billion kelvin. At this temperature, matter within the star starts to emit photons that are so dangerously energetic they can create pairs of electrons and positrons when they collide. The result is a runaway reaction that obliterates the star in a colossal explosion.

The first suspected sighting of such a 'pair-instability super-nova' came in 2007, when a brilliant and exceptionally long-lasting stellar explosion was observed, suggesting the existence of a star far bigger than had previously been thought possible.

During a supernova, stellar temperatures can briefly rise far above 6 billion kelvin. In 1987, a star was seen exploding in the Large Magellanic Cloud, a satellite galaxy of our own Milky Way some 160,000 light years away from us. Neutrinos from its heart detected on Earth revealed an internal temperature of about 200 billion kelvin.

That's nothing, though, compared with whatever produces a gamma-ray burst. These brief flashes of ultra-high-energy light are spotted once or twice a day by specially tuned telescopes. Gamma-ray bursts are thought to mark the birth of black holes, either when a giant star's core collapses or when two ultra-dense neutron stars collide. Somehow the gravitational energy is turned into a tight beam of gamma rays and other radiation. While the details of this process are currently unknown, it must involve a fireball of relativistic particles heated to something in the region of a trillion kelvin ($10^{12}$ K).

Closer to home is a place that is even hotter: not a natural inferno, but a detector cavity 100 metres or so beneath the generally temperate outskirts of Geneva. There, between 8 November and 6 December 2010, nuclei of lead atoms were smashed together for the first time at CERN's Large Hadron Collider in an attempt to mimic some of the universe's opening moments. The result was the highest temperatures ever recorded on Earth, a subatomic fireball registering several trillion kelvin.

That experiment gives us a clue to where the universe's extreme of heat lies. Not in the here and now, but the way back when. Looking into the heart of the big bang, the singularity of temperature and density in which our universe began, the maximum temperature is just a matter of how many zeros you can write before our understanding of physics breaks down. That's probably somewhere in the region of 32.

## The coldest thing in the universe

Space itself is neither hot nor cold. In the absence of stuff with thermal vibrations, temperature has no meaning. But there are plenty of cold things *in* space.

In our solar system, the coldest known spot is quite close by. In 2009, NASA's Lunar Reconnaissance Orbiter found permanently shadowed craters near the south pole of the moon that were at only 33 kelvin (-240 °C). By comparison, the frigid haze around Pluto is a relatively balmy 70 kelvin (-203 °C), as measured by New Horizons probe in 2015. As exploration continues and measurements improve, that record is likely to pass to some moon or dwarf planet much further from the sun, perhaps with its own sheltered and frigid craters.

Beyond our solar system there are certain to be some even chillier rocks, and the coldest of all these lonely wanderers are likely to be found in intergalactic voids. Warmed only by the weak microwave afterglow of the big bang and a glimmer of distant starlight, their temperature would be no more than 3 K (-270 °C).

Since the 2.7 K microwave background bathes the entire universe, you might imagine that nothing could be colder than this. Not so. A gas cloud called the Boomerang nebula, 5000 light years away, has a temperature of only 1 K. The nebula is expanding rapidly, which actively cools its gas in the same way that expansion chills the coolant in a domestic refrigerator or aircon unit.

Whether the Boomerang retains its status as the coldest known natural object remains to be seen, but this is one area in which humans have no trouble outdoing nature. In 2003, a cloud of sodium atoms in a lab at the Massachusetts Institute of Technology was chilled to 0.45 nanokelvin, less than half-a-billionth of a degree above absolute zero – far colder than any temperature the wider universe seems to have a use for.

## The densest thing in the universe

At the modest temperatures and pressures of Earth's surface, the densest known material is the metallic element osmium, which packs 22 grams into 1 cubic centimetre, or more than 100 grams into a teaspoonful. Even osmium is full of fluff, however, in the form of electron clouds that separate the dense atomic nuclei. Although rarefied, these clouds are robust, and even the immense pressures deep within the planet can only compress solid matter to a modest degree.

Far greater pressure is found within the collapsed core of a giant star, a remnant we know as a neutron star. There, matter is in some exotic and ultra-dense form – most probably neutrons, and possibly a few protons and electrons, packed cheek-by-jowl. One cubic metre of 'neutronium' matter from the centre of a neutron star could have a mass of up to $10^{18}$ kilograms, or a million billion tonnes.

An even denser hypothetical material may yet exist in the cores of neutron stars: quark matter, in which protons and neutrons dissolve into their constituent particles. The latest evidence is against it, though. In 2010, astronomers spotted two neutron stars so heavy that they would probably squeeze a quark-matter core into oblivion. The clues to what really lies at the heart of a neutron star may come through studying starquakes, the juddering explosions of energy that happen when the crust of a neutron star ruptures.

Neutronium, or perhaps quark matter, may be the densest form of matter in the cosmos, but it is probably not what the densest object is made of. Compress a neutron star even further, and it will turn into a black hole. Not that all black holes are particularly dense: in fact the big ones, as measured by their event horizons, are quite tenuous. A supermassive black hole in the nearby galaxy M87 has a mass 6.4 billion times that of our sun but a density of only 0.37 kilograms per cubic metre, making it lighter than air.

On the other hand, the smallest known black hole – a minnow called XTE J1650-500 – is only 3.8 times the mass of the sun, but its density is just over $10^{18}$ kilograms per cubic metre. Find one of these warps in space–time that is just a little smaller, and it will overtake neutronium in the density stakes.

Microscopic black holes might also have been forged during the big bang, when quantum fluctuations in a hugely dense universe could have led to regions so dense that they collapsed. Such micro-holes might yet reveal themselves in sudden bursts of radiation: if so, this could give us an insight into the scale of quantum fluctuations in the nascent universe, and perhaps what processes actually drove the big bang.

Inside a black hole's event horizon things get even stranger. The theory of relativity tells us that all that mass is squeezed down to a mathematical point of infinite density – though the theory almost certainly breaks down at such extremes as quantum effects begin to scramble space–time. Here, where gravity meets the quantum world, is the great frontier of fundamental physics. It is by considering such extremes as black-hole singularities that theoreticians hope to understand the most profound basis of reality.

Does a black hole's heart conceal a fuzzball of wobbling strings? Or a space-sucking quantum wormhole? We don't know, although back-of-the-envelope calculations suggest an upper limit on its density of $5 \times 10^{96}$ kilograms per cubic metre, called the Planck density. The densest thing in the universe can probably be no denser than that – whatever it actually is.

## The fastest thing in the universe

Speed is relative. There is no absolute standard for 'stationary' in the universe. Perhaps the nearest thing is the all-pervasive cosmic microwave background radiation (CMB). Its Doppler shift across the sky – blue in one direction, red in the other – reveals that, relative to the CMB, the solar system is rattling along at 600 kilometres per second. Microwaves are rather insubstantial, though, so we don't feel the wind in our hair.

Distant galaxies are also moving at quite a rate. Space is expanding everywhere: the more space you are looking through the faster the galaxies you see are moving away from us. Far enough off, galaxies are effectively retreating faster than light speed, which means we can never see them because their radiation can't reach us.

While such inaccessible extremes may have abstract appeal, speed becomes much more interesting if you are moving fast relative to some large object nearby – something you can see whoosh past your windows, or something you might just crash into.

Within our solar system, Mercury, the messenger of the gods, is the fastest-moving planet, with an orbital speed of about 48 kilometres per second. Earth manages only about 30 km/s. In 1976, Mercury was outpaced for the first time by a human artefact, the *Helios 2* solar probe, which reached more than 70 km/s as it whizzed by the sun. Sun-grazing comets that swoop in from the outer solar system trump both, skimming past the solar surface at up to 600 km/s. Speed does not guarantee escape: a few hit the sun and are swallowed.

The outer reaches of the Milky Way are also home to some oddly busy bodies: 'hypervelocity stars' speeding past the rest of the galaxy at up to 850 km/s. The theory is that they were flung out in a close encounter with the giant black hole in our galaxy's centre. Black holes make particularly effective cosmic slingshots because of their peerlessly powerful gravity. Some also create magnetic tornadoes that squirt out tenuous jets of matter at more than 99 per cent of the speed of light.

The spinning neutron stars we know as pulsars also perform high-speed magnetic magic. Pulsars can rotate up to 1000 times a second, which means their surfaces move at up to 20 per cent of the speed of light. Far enough away from the surface, the magnetic field projected by the pulsar can even move faster than light. That is not in conflict with the laws of physics as the magnetic field carries no energy or information. These superfast fields are perhaps the source of the powerful, regular pulses of radiation pulsars send our way.

Even solid objects can approach light speed, with the aid of a black hole's gravity. At a hole's event horizon, a single rock will simply disappear without a splash, but two rocks on different trajectories could collide with one another. According to calculations posted online in 2010 by Tomohiro Harada at the University of Tokyo, and his colleague Masashi Kimura, the rotation of the black hole whips up a whirlpool in the surrounding space and increases the maximum collision speed. The upshot is that somewhere in the universe, two rocks caught in the grip of a rapidly spinning black hole could be hurtling towards one another at close to the speed of light.

## The brightest thing in the universe

Everyday units are far too small to cope tidily with the brilliance of the cosmos. Instead, astronomers use the sun, and its dazzling light output of $4 \times 10^{26}$ watts, as a standard lamp.

The sun is in fact an above-average star in terms of bright-ness, but some stellar show-offs outshine it by far. The most luminous example clearly visible to the naked eye is Epsilon Orionis, the middle star of Orion's Belt. This blue supergiant is 1300 light years away and 400,000 times as bright as the sun. Much further away within our galaxy, or obscured by dust, are yet brighter stars such as the unstable Eta Carinae, which pumps out as much light as 5 million suns.

In July 2010, astronomers found a new record breaker. R136a1 is a star in the Large Magellanic Cloud that is as bright as almost 9 million suns. With a mass estimated to be 250 times that of the sun, this freakish body is heavier than anyone thought possible, at least for a star made from the kind of chemical mixture available in the gas of the Milky Way and its neighbours. Could it be built from an almost pure source of hydrogen and helium gas that had somehow survived uncontaminated since the early days of the universe, or is there something wrong with our theories of stellar structure?

Some massive stars burn brighter still – but only for a few weeks and at the cost of their lives. A supernova called SN 2005ap, in a galaxy 4.7 billion light years away, qualifies as the brightest stellar explosion on record, peaking at about 100 billion suns.

Gamma-ray bursts emit even more energy than a super-nova, and they can pack it into a matter of seconds rather than spreading it over several weeks. A burst can make even our solar unit seem absurdly feeble: its luminosity can equal more than $10^{18}$ suns.

If such explosions seem unsatisfyingly transient, then the brightest steady lights in the universe are quasars, in which a massive black hole feeds on a copious supply of stars and gas. As this doomed material spirals inwards it becomes white hot, and it can shine with the light of more than thirty trillion suns.

## The roundest thing in the universe

In medieval cosmology, the universe was a nested series of perfect crystal spheres that carried the sun, moon, planets and stars. We now know that space is rather messier, but does it hold anything to mirror that vision of spherical perfection?

Planets themselves are pulled into fairly tidy spheroidal shapes by the force of their own gravity. The most prominent of Earth's bumps and its deepest wrinkles, from Mount Everest to the Mariana trench, point in or out by less than 0.2 per cent of the planet's radius. If it weren't for the slightly squashed shape caused by Earth's daily rotation – pulled in at the poles, bulging at the midriff – our home would make a good cosmic pool ball.

Yet Earth is positively craggy compared with neutron stars. Their huge density results in a surface gravity something like 200 billion times as strong as Earth's. That is enough to flatten out all but the slightest irregularity: a neutron star's Everest would probably be no more than 5 millimetres high. As these stars are typically 10 to 15 kilometres across, that Himalayan height is less than one part in a million of the stellar radius.

For a period of 16 months during 2004 and 2005, we launched our own balls into space that rivalled neutron stars for roundness. *Gravity Probe B* was a satellite designed to look for distortions in space–time created by our planet's great mass, which are predicted by Einstein's general theory of relativity. One of these is an effect called frame-dragging, in which space is dragged around with the rotation of Earth. *Gravity Probe B* used four gyroscopes based on small spheres of quartz polished so thoroughly that they have no irregularities larger than 0.4 parts in a million.

Relativity happens to offer us something rounder even than that probe's spheres. A black hole's event horizon marks the region from which no light can escape to reach the eye of a distant observer. It isn't exactly a surface: you couldn't run a hand over it and marvel at its new-shave smoothness. But soon astronomers may be able to discern an image of the supermassive black holes at the centre of the Milky Way using a planet-wide array of radio observatories called the Event Horizon Telescope. Techniques like this could eventually give us a sharp picture of these pseudosurfaces which are perhaps the nearest thing in nature to perfect roundness.

Observing matter falling in to an event horizon could be a sterner test of Einstein. If we see shreds of gas on orbits slightly different from the predictions of relativity, we may need a new theory of gravity. And of course if black holes turn out not to have the expected event horizon, that would be a shocker.

## The darkest thing in the universe

Galaxies are supposed to be glittering jewels, studded with billions of bright stars and glowing nebulae. Not so Segue 1, the dark horse of the galactic neighbourhood. Segue 1 is only 75,000 light years away, making it a near neighbour of the Milky Way, yet it remained undiscovered until 2006 because its total light emission is only 300 times that of our sun.

That is odd. Segue 1's few stars are moving around quite fast, so its gravity must be reasonably strong, implying that it contains at least a million solar masses of matter. Very little of that can be accounted for by visible stars and gas, suggesting that almost all of it must be exotic dark matter.

Studying dwarf galaxies like Segue 1 could tell us more about dark matter. For example, if the cores of these galaxies are less dense than predicted by the standard assumptions of how cold dark matter should behave, it could mean that the stuff is warm, or prone to self-destruct, or made from ultra-light particles that are inherently fuzzy.

Even better would be finding a 'dark star' – a cool, fat blob of gas gently warmed from within by decaying dark matter. Such beasts are thought to have existed in the very early universe, and there may still be a few around today, but none has yet been spotted.

Meanwhile, CERN's Large Hadron Collider is being used to hunt for possible dark-matter particles, so perhaps the hottest thing on Earth will soon illuminate the dimmest thing in space.

## The atoms at infinity degrees Celsius – and beyond

*Going into negative temperatures sounds cool, but turns out to be the opposite.* **Jacob Aron** *watches the mercury rise.*

Nothing is colder than absolute zero, but there's a whole world beyond that point. In 2013, scientists put atoms in a state where they could take on a range of negative absolute temperatures. This could deepen our understanding of sub-atomic particles – and perhaps even mysterious dark energy.

Although we're used to talking about negative temperatures, such as -10 °C, all temperatures on an ordinary thermometer are positive when measured in kelvin. On this scientific scale, which starts at absolute zero (-273.15 °C), temperature is determined by the kinetic energy of particles. So a gas of slow particles is colder than a gas of fast-moving ones, and absolute zero corresponds to the point at which particles stop moving completely, which is why nothing can be colder.

That does not tell the whole story, though. Temperature also depends on how the particle energies are distributed within a sample, which sets their entropy, a thermodynamic property.

At absolute zero, all particles have minimum energy, and zero entropy. As they heat up, some particles gain lots of energy but most just a small amount. This increase in the energy spread means higher entropy – but only until the particles are evenly distributed between all possible energies, corresponding to maximum entropy.

This is the maximum possible positive temperature, but you can still increase the particles' energy. However, from now on, more and more particles will belong to the high energy group, so entropy will go down with further energy hikes, a sign that the particles have flipped into the realm of negative temperature, which, bizarrely, is hotter than its positive counterpart.

'The temperature scale as we know it starts at zero and goes up to infinity, but it doesn't stop there,' says Ulrich Schneider of the Ludwig Maximilian University of Munich. The resulting thermometer is mind-bending. It starts at zero, ramps up to infinity, then jumps to the negative side of infinity before increasing through the negative scale to zero.

Previously researchers entered the negative realm by flipping a magnetic field so that an array of mostly aligned atomic nuclei become mostly misaligned. But this only offered two discrete energy states – one negative, one positive. To create a system with a range of negative temperatures, Schneider and his colleagues used moving atoms, following a recipe first proposed in 2005.

They began by cooling atoms to a fraction above absolute zero and placing them in a vacuum. Lasers then placed the atoms along the curve of an energy 'valley', with the majority of the atoms in lower-energy states. Next, the researchers turned this positive system negative by adjusting the lasers to change the atoms' energy levels so that the majority of the particles were high energy. The atoms can't lose energy and 'roll down' the hill because doing so would mean first increasing their kinetic energy, which is not possible because they are in a vacuum and there is no outside energy source. So this flipped the valley into an energy hill, producing the inverse energy distribution that characterises negative temperatures. What's more, unlike in the magnetic system, lasers can set the hill at different heights, giving a range of negative temperatures.

Cold atoms are already used to simulate the interactions of subatomic particles, including quarks. Schneider's set-up could create simulations not possible with positive temperatures. Negative temperature may also have implications for cosmology. Dark energy, thought to explain the accelerating expansion of the universe, exerts negative pressure, which suggests it might have negative temperature.

## The giants dancing the slowest ballet on Earth

*Plate tectonics is a slow-grind drama with some dramatic plot twists.* **Stephen Battersby** *turns the clocks forward 250 million years to see how Earth looks.*

Asia is torn in two. The Atlantic and Pacific oceans are swallowed. Where once there were beaches, great mountain ranges judder into the skies, fusing together a scatter of separate land masses into one mighty new supercontinent. Call it . . . Aurica.

That's what João Duarte calls it, anyway. A geoscientist at the University of Lisbon, he has his own distinct vision of how Earth may look 250 million years from now. He joins a band of fortune tellers gazing into the distant future, all with different ideas about how and where the next supercontinent will form, and what cataclysms might strike along the way.

The answer will determine Earth's future climate and prospects for sustaining life. But getting it right requires grappling with a machine whose workings we still understand only imperfectly: that of plate tectonics.

Earth's surface is clad in rigid rock plates – together called the lithosphere – formed of surface crustal rock laminated on to hard cold mantle rocks. Given their rigidity, it is surprising that these plates don't simply lock together, unmoving. And indeed, until about 50 years ago geologists thought that Earth's land masses were fixed, despite German geophysicist Alfred Wegener having proposed the idea of continental drift in 1915.

The creation and destruction of ocean basins makes plate motion possible. Plates move apart at mid-ocean ridges, where molten rock rises and cools to form hard, dense basalt. They move together at subduction zones, where old ocean lithosphere plunges under a neighbouring plate. As it penetrates the warmer, softer mantle beneath, it causes earthquakes and feeds volcanoes.

Magnetic signals recorded in sea-floor rocks, and chemical traces from the roots of ancient mountain ranges, tell us how continental drift has changed the face of Earth. They point clearly to a time 180 million years ago when all today's continents were stuck together in one vast land mass centred roughly where present-day Africa is: the supercontinent Pangaea, from the Ancient Greek for 'all of Earth'.

## Cycling continents

We know that Pangaea came together about 330 million years ago. Before that, some consider a relatively short-lived gathering of continents near the South Pole to be another supercontinent, named Pannotia or Greater Gondwana. Another supercontinent, Rodinia, probably dominated the planet between about 1.2 billion and 700 million years ago. And about 2 billion years back it is thought there was another, known as Nuna or Columbia.

What has been will be. 'For more than 20 years we have recognised that Pangaea was just the latest in a series of supercontinents,' says Brendan Murphy at St Francis Xavier University in Antigonish, Canada. 'That implies there will be another one in the future.'

What's unclear is how that vast land mass will form. One model simply projects what's happening today into the future. The great split that broke Pangaea apart is still growing, and the two biggest land masses on Earth, Africa-Eurasia on the one hand and the Americas on the other, are on the move. The Atlantic is spreading as new rock wells up at its mid-ocean ridge, while the Pacific is shrinking, consumed by the subduction zones that surround it, the famous ring of fire.

'If you simply run plate tectonics forward in time, you would see the Pacific close and the Atlantic open,' says Mark Behn at Woods Hole Oceanographic Institution in Massachusetts. In about 250 million years a new supercontinent, Novopangaea, would form on the opposite side of Earth from the original Pangaea, as the Americas and Asia crunch in around northbound Australia.

But it may not be that simple. 'You can get to 50 million years by projecting present-day motions,' says Christopher Scotese at Northwestern University in Evanston, Illinois. 'To go any further, you have to understand why plates are moving, what are the driving forces, what are the rules of plate tectonics.' That's something we are still struggling to do.

Since 1982, Scotese has been making maps of past and future Earth using various rules of thumb. The most important rule is widely accepted: plate tectonics is driven mainly by the pull of sinking slabs at subduction zones, with a smaller push from new rock forming at mid-ocean ridges. Work out the layout of subduction zones and ridges at any time, and you can begin to see how the continents should be tugged and nudged around.

But three kinds of cataclysmic event can change the course of this smooth voyage. A subduction zone can swallow a spreading ocean ridge, as is happening today off North America's west coast, where the Juan de Fuca ridge is slowly being consumed. Or a pair of unsinkable continents can collide, snuffing out subduction in between and forcing a great mountain range, as India and Eurasia have done to build the Himalayas. The third possible cataclysm is much harder to fathom or predict. 'The fundamental question is how you start new subduction zones,' says Behn.

This has to happen somehow, or all existing subduction zones would eventually be killed by continental collisions, and plate tectonics would cease altogether. In 2008, Behn and his colleague Paul Silver suggested that if a supercontinent forms simply by closing the Pacific, destroying all the main subduction zones around it, plate tectonics could shut down for a long time. They suggest something similar may have happened in the past.

But even if that is true, evidence that plate tectonics has gone on for almost all of Earth's history, with many cycles of supercontinent formation and destruction, indicates that new subduction must start eventually, somewhere.

The most likely spot is at passive margins, for example on the Atlantic coasts of Europe, Africa and the Americas. These are places where old oceanic lithosphere, spreading out from mid-ocean ridges, meets continental crust. The oceanic lithosphere has had time to cool since formation and become denser than the rock beneath, so it wants to sink.

But it can't. Old, cold lithosphere rocks are hard to crack. Even the weight of kilometres-deep river sediments washed on to the passive margin from the continents isn't enough on its own. The weakening effect of water seeping into the rocks may help, but probably not enough to crack those passive margins.

In the 1980s, Scotese suggested that subduction is catching. While it is difficult to break old ocean lithosphere from scratch, 'a nick will localise stresses and tear more easily', he says. There are already two small subduction zones in the western Atlantic: the Lesser Antilles volcanic arc near the Caribbean, and the Scotia arc in the far south off Tierra del Fuego. These both look as though they have sneaked in from old subduction zones in the Pacific Ocean. Scotese suggests that eventually they will spread south and north, joining up to form a long subduction zone up the east coast of the Americas. In his projection, this will eat the Atlantic's mid-ocean ridge about 100 million years from now, and the Atlantic will start to close again. After 250 million years, the Americas will have collided with an already merged Africa and Eurasia – as will Australia and most of Antarctica – to form what Scotese calls Pangaea Proxima.

Others have come to a similar conclusion by looking at the geological record, which shows that oceans periodically open and close in something known as the Wilson cycle. In the 1980s, Thomas Worsley and Damian Nance at Ohio University suggested that the next supercontinent might form more or less in the way it split up, by closing the Atlantic.

## Old and crusty

In 2012, Ross Mitchell, then at Yale University, and his group mapped out a third route. The shifting of mass associated with the formation of a supercontinent affects Earth's rotation, changing its spin axis relative to the solid body of the planet. By looking at the orientation of magnetic crystals in rocks that cooled around the time that different supercontinents existed, the team showed that Rodinia formed about 90 degrees in latitude away from the position of Nuna, and Pangaea about 90 degrees from Rodinia. Mitchell and his colleagues predict that the same thing will happen again, meaning the next supercontinent should form somewhere near the North Pole, as Asia and North America crunch together. They call the result Amasia.

Some support for this view came in 2016, when Masaki Yoshida at the Japan Agency for Marine-Earth Science and Technology in Yokosuka published a numerical simulation of mantle motion. It shows continents converging near the North Pole, guided there partly by plumes of hot, rising mantle rocks that help to keep the Pacific open in the south while it closes in the north. This makes Amasia relatively straggly, with the Americas forming a huge promontory and Antarctica remaining aloof, unlike the compact form of the original Pangaea and the other imagined supercontinents of the future.

Duarte thinks all these models have problems. Amasia and Novopangaea would both be surrounded by large areas of ocean crust that is more than 400 million years old, which he finds implausible. In 2008, Dwight Bradley at the US Geological Survey in Anchorage, Alaska, looked at rocks around ancient passive margins and found the oldest were about 180 million years old on average, and much less than that in recent ages. Hardly any lasted 400 million years. Duarte thinks this is no coincidence. 'Somehow plates in Atlantic-type oceans may have to start subducting after about 200 million years,' he says.

Scotese's Pangaea Proxima does not have the old-crust problem: the Pacific could in theory stay open for many hundreds of millions of years with new crust constantly being generated and destroyed. But Duarte considers this improbable too, because ridges such as the Juan de Fuca are already being subducted. 'It may not be very likely that new ones form in the middle of the oceanic plates where they are cold and strong,' he says.

Duarte agrees with Scotese that subduction may spread like a virus, a process he calls invasion. He has found evidence that subduction is beginning to invade the Atlantic's eastern margin, off the coast of Portugal, where forces generated by the remnants of an ancient subduction in the Mediterranean are helping to create new faults in the ocean floor.

In Duarte's model of Earth's future, published last year, subduction will spread along both sides of the Atlantic within a few tens of millions of years, and the ocean will begin to close. But the Pacific will keep on closing, too, meaning something else has to give. That something is Asia. A rift cuts across the continent, from the Indian Ocean up to the Arctic, as the Himalayan Plateau collapses under its own weight. A new ocean opens up, and the eventual outcome is a new supercontinent with the two halves of Asia on the outside and American and Australia at its core – hence Aurica.

Nice try, says Scotese. 'Trying to close the Atlantic and Pacific – I think that's original.' But suggesting a subduction zone on the eastern as well as the western side of the Atlantic actually makes things more difficult, he says, because that could preserve the mid-Atlantic ridge. 'To close the ocean you have to subduct the ridge – but if you have subduction all around the ocean, the ridge can stay in the middle and supply crust to both sides.'

The proponents of each idea are keen to stress that the

future is uncertain, and that their own model is just one option (although of course the most likely one). Whoever is right, our distant cousins will have to adapt to a strangely shaped world, and will in turn be shaped by it. 'The whole Earth system is ultimately controlled by plate tectonics,' says Scotese. As continents move through different climate zones, they cause new problems for existing life forms and create opportunities for others. Extreme volcanism can also cause or at least contribute to mass extinctions, as in the vast outpouring of lava that formed the Deccan Traps in India about 65 million years ago. This changed the global climate and may have put the dinosaurs under serious stress, before a meteorite provided the knockout.

## Driving forces

Climate may differ wildly between different supercontinent scenarios, affecting Earth's habitability. Amasia, near the North Pole, might gather a massive ice cap. Novopangaea could be similar to original Pangaea, which may have seen extremes of weather, with a vast interior desert and a seasonal 'mega-monsoon'. Or if plate tectonics were to shut down for a while, that would radically affect the atmosphere. Volcanoes would cease to pump out carbon dioxide, and the planet might enter a severe ice age. We could model the climate on each hypothetical supercontinent – 'but that is building a house of cards then building balconies on your house of cards,' says Scotese.

Earth's future depends on what forces really drive the motion of the continents, and that remains the real unknown. The only way we can cut through the profusion of possibilities is to make plate tectonics more quantitative, says Duarte. 'To make substantial progress we need more observations of

Earth's interior,' he says. 'For example, there are barely any permanent seismometers on the bottom of the oceans.' We could also use neutrinos from the sun to probe the planet's interior, he suggests. Certainly, more powerful models that can capture geochemical processes on scales large and small are needed.

So trying to gaze into Earth's distant future may help us to understand the inner workings of our complex and opaque world today. But important though that is, it's not the real motivation for sketching out the next supercontinent, 10 million generations removed. 'It's for fun,' says Duarte.

## The weirdest stuff in the world

*The quantum world is strange and not always small. In the coldest labs in the universe, **Michael Brooks** discovers bucketfuls of liquid flowing uphill and solids passing through one another.*

For centuries, con artists have convinced the masses that it is possible to defy gravity or walk through walls. Victorian audiences gasped at tricks of levitation involving crinolined ladies hovering over tables. Even before then, fraudsters and deluded inventors were proudly displaying perpetual-motion machines that could do impossible things, such as make liquids flow uphill without consuming energy. Today, magicians still make solid rings pass through each other and become interlinked – or so it appears. But these are all cheap tricks compared with what the real world has to offer.

Cool a piece of metal or a bucket of helium to near absolute zero and, in the right conditions, you will see the metal levitating above a magnet, liquid helium flowing up the walls of its container or solids passing through each other. 'We love to observe these phenomena in the lab,' says Ed Hinds of Imperial College London.

This weirdness is not mere entertainment, though. From these strange phenomena we can tease out all of chemistry and biology, find deliverance from our energy crisis and perhaps even unveil the ultimate nature of the universe. Welcome to the world of superstuff.

This world is a cold one. It only exists within a few degrees of absolute zero, the lowest temperature possible. Though you might think very little would happen in such a frozen place, nothing could be further from the truth. This is a wild, almost surreal world, worthy of Lewis Carroll. One way to cross its threshold is to cool liquid helium to just above 2 kelvin. The first thing you might notice is that you can set the helium rotating, and it will just keep on spinning. That's because it is now a 'superfluid', a liquid state with no viscosity.

Another interesting property of a superfluid is that it will flow up the walls of its container. Lift a bucketful of superfluid helium out of a vat of the stuff, and it will flow up the sides of the bucket, over the lip and down the outside, rejoining the fluid it was taken from. Though fascinating to watch, such gravity-defying antics are perhaps not terribly useful. Of far more practical value are the strange thermal properties of superfluid helium.

Take a normal liquid out of the refrigerator and you find it warms up. With a superfluid, though, the usual rules no longer apply. Researchers working at the Large Hadron Collider at CERN, near Geneva, use this property to help accelerate beams of protons. They pipe 120 tonnes of superfluid helium around the accelerator's 27-kilometre circumference to cool the thousands of magnets that guide the particle beams. Normal liquid helium would warm up considerably if used in this way, but the extraordinary thermal properties of the superfluid version means its temperature rises by less than 0.1 kelvin for every kilometre of the beam ring. Without superfluids, it would have been impossible to build the machine that many physicists hope will reveal the innermost secrets of the universe's forces and building blocks.

The LHC magnets have super-properties themselves. They are made from the superfluid's solid cousin, the superconductor. At temperatures approaching zero kelvin, many metals lose all resistance to electricity. This is not just a gradual reduction in resistance, but a dramatic drop at a specific temperature. It happens at a different temperature for each metal, and it unleashes a powerful phenomenon.

For a start, very little power is needed to make superconductors carry huge currents, which means they can generate intense magnetic fields – hence their presence at the LHC. And just as a superfluid set rotating will keep rotating forever, so an electric current in a superconducting circuit will never fade away. That makes superconductors ideal for transporting energy, or storing it.

The cables used to transmit electricity from generators to homes lose around 10 per cent of the energy they carry as heat, due to their electrical resistance. Superconducting cables would lose none. Storing energy in a superconductor could be an even more attractive prospect. Renewable energy sources such as solar, wind or wave power generate energy at an unpredictable rate. If superconductors could be used to store excess power these sources happen to produce when demand is low, the world's energy problems would be vastly reduced.

We are already putting superconductors to work. In China and Japan, experimental trains use another feature of the superconducting world: the Meissner effect. Release a piece of superconductor above a magnet and it will hover above it rather than fall. That's because the magnet induces currents in the superconductor that create their own magnetic field in opposition to the magnet's field. The mutual repulsion keeps the superconductor in the air. Put a train atop a superconductor and you have the basis of a levitating, friction-free transport system. Such 'maglev' trains do not use metal superconductors because it is too expensive to keep metals cooled to a few kelvin; instead they use ceramics that can superconduct at much higher temperatures, which makes them much easier and cheaper to cool using liquid nitrogen.

## A tale of two particles

These are strange behaviours indeed, so what explains them? Both superfluidity and superconductivity are products of the quantum world. Imagine you have two identical particles, and you swap their positions. The physical system looks exactly the same, and responds to an experiment exactly as before. However, quantum theory records the swap by multiplying their quantum state by a 'phase factor'. Switching the particles again brings in the phase factor a second time, but the particles are in their original position and so everything returns to its original state. 'Since switching the particles twice brings you back to where you were, multiplying by this phase twice must do nothing at all,' says John Baez at the Centre for Quantum Technologies in Singapore. This means that squaring the phase must give 1, which in turn means that the phase itself can be equal to 1 or -1.

This is more than a mathematical trick: it leads nature to divide into two. According to quantum mechanics, a particle can exist in many places at once and move in more than one direction at a time. Last century, theorists showed that the physical properties of a quantum object depend on summing together all these possibilities to give the probability of finding the object in a certain state.

There are two outcomes of such a sum, one where the phase factor is 1 and one where it is -1. These numbers represent two types of particles, known as bosons and fermions.

The difference between them becomes clear at low temperatures. That is because when you take away all thermal energy, as you do near absolute zero, there aren't many different energy states available. The only possibilities to put into quantum theory's equations come from swapping the positions of the particles.

Swapping bosons introduces a phase change of 1. Using the equations to work out the physical properties of bosons, you find that their states add together in a straightforward way, and that this means there is a high probability of finding indistinguishable bosons in the same quantum state. Simply put, bosons like to socialise.

In 1924, Albert Einstein and Satyendra Bose suggested that at low-enough temperatures, the body of indistinguishable bosons would effectively coalesce together into what looks and behaves like a single object, now known as a Bose–Einstein condensate, or BEC.

Helium atoms are bosons, and their formation into a BEC is what gives rise to superfluidity. You can think of the helium BEC as a giant atom in its lowest possible quantum energy state. Its strange properties derive from this.

The lack of viscosity, for instance, comes from the fact that there is a huge gap in energy between this lowest state and the next energy state. Viscosity is just the dissipation of energy due to friction, but since the BEC is in its lowest state already, there is no way for it to lose energy – and thus it has no viscosity. Only by adding lots of energy can you break a liquid out of the superfluid state.

If you physically lift a portion of the superatom, it acquires more gravitational potential energy than the rest. This is not a sustainable equilibrium for the superfluid. Instead, the superfluid will flow up and out of its container to pull itself all back to one place. Superconductors are also BECs. Here, though, there is a complication because electrons, the particles responsible for electrical conduction, are fermions.

Fermions are loners. Swap them around and, as with swapping your left and right hand, things don't quite look the same. Mathematically, this action introduces a phase change of -1 into the equation that describes their properties. The upshot is that when it comes to summing up all the states, you get zero. There is zero probability of finding them in the same quantum state.

We should be glad of this: it is the reason for our existence. The whole of chemistry stems from this principle that identical fermions cannot be in the same quantum state. It forces an atom's electrons to occupy positions further and further away from the nucleus. This leaves them with only a weak attraction to the protons at the centre, and thus free to engage in bonding and other chemical activities. Without that minus sign introduced as electrons swap positions, there would be no stars, planets or life.

So how do the electrons in superconductors form into BECs? In 1956, Leon Cooper showed how electrons moving through a metal can bind together in pairs and acquire the characteristics of a boson. If all the electrons in a metal crystal form into such Cooper pairs, these bosons will come together to form, as in superfluid helium, one giant particle – a BEC.

The main consequence of this is a total lack of electrical resistance. In normal metals, resistance arises from electrons bumping into the metal ions bouncing around. But once a metal becomes a superconductor, the electron-pair condensate is in its lowest possible state. That means it cannot dissipate energy and, once the Cooper pairs are made to flow in an electrical current, they simply keep flowing. The only way to disturb superconductivity without raising the temperature is to add energy another way, for example by applying a sufficiently strong magnetic field.

Though superfluids and superconductors are bizarre enough, they are not the limit of the quantum world's weirdness, it seems. 'There is yet another level of complexity,' says Ed Hinds. That complexity comes into play below 1 kelvin and at more than 25 times Earth's atmospheric pressure, when helium becomes a solid. This form of helium plays havoc with our notions of solidity. Get the conditions right and you can make solids pass through each other like ghosts walking through walls.

Such an effect was first observed in 2004 by Moses Chan and Eunseong Kim at Penn State University in University Park, Pennsylvania. They set up solid helium in a vat that could rapidly rotate back and forth, inducing oscillations in the solid helium. They observed a resonant vibrational frequency which they interpreted as indicating that there were two solids in the vat, which were passing through each other.

Admittedly the two solids do not fit our usual definitions. One was made up of 'vacancies', created when helium atoms shake free of the lattice that forms solid helium. The gaps left behind have all the properties of a real particle – they are so like real particles, in fact, that their quantum states can lock together to form a BEC. The solid helium is also a BEC, and it is these two condensates that pass through each other.

Chan and Kim's observation is still somewhat controversial; some researchers think there is a more prosaic explanation to do with deformations and defects in the helium lattice. 'There is a lot of activity, several theory notions and experiments of interest, but no real agreement,' says Robert Hallock of the University of Massachusetts at Amherst.

Nonetheless, even the fact that it might be possible to create solids that aren't really solid shows just how odd superstuff can get. And it's all because the world has a fundamental distinction at its heart. Everything, from human beings to weird low-temperature phenomena like liquids that defy gravity, stems from the fact that there are two kinds of particles: those that like to socialise, and those that don't. Sound familiar? Perhaps the quantum world isn't that different from us after all.

# 7 A Practical Guide to Blowing Your Mind

According to legend, the Gordian knot securing an ox-cart in the kingdom of Phrygia was one so fiendishly complex that nobody could untie it: an oracle proclaimed that the person who did would rule over all Asia. Greater acclaim still might be in store for the person who can unravel the workings of the mind, a tangled mass of neurons no less complicated.

Nonetheless, there's no shortage of people willing to tug at a loose thread to see what happens. From scientists to self-experimenters, this chapter chronicles some of pioneering work going on to shed light on the mysteries of the human mind, and how we can intervene to prevent disease, heal trauma and pursue healthier, happier lives.

You'll learn how holes drilled into the skull can create a peephole to its inner workings, the superpowers your mind offers and how to boost them, the party drugs that grew up and got proper jobs helping to save lives and the enduring mystery of what anaesthesia does to the brain.

In the end, it was Alexander the Great who finally loosed the Gordian knot: by slicing through it with his sword. An acceptable solution for freeing ox-carts maybe, but please do take more care with your mind.

# How psychedelic drugs are rebuilding broken brains

*The healing powers of illegal drugs like MDMA and psilocybin are finally living up to the hype – and they are already transforming our view of mental illness, says **Sam Wong**.*

He wasn't the first person to say it, and he probably won't be the last, but Tom Insel's accusation carried extra weight thanks to his job title: director of the US National Institute of Mental Health. Towards the end of his 13-year tenure, Insel began publicly criticising his own organisation, and psychiatry in general, for its failure to help people with mental illness. 'There are great examples in other areas of medicine where we've seen innovation really make a difference,' says Insel. 'Not so much for patients with schizophrenia, post-traumatic stress disorder or depression.'

It's hard to argue. Mental illness has reached crisis proportions, yet we still have no clear links between psychiatric diagnoses and what's going on in the brain – and no effective new classes of drugs. There is one group of compounds that shows promise. They seem to be capable of alleviating symptoms for long periods, in some cases with just a single dose. The catch is that these substances, known as psychedelics, have been outlawed for decades.

A psychedelic renaissance has been feted many times, without ever delivering on the high hopes. But this time feels different. Now there is a growing band of respected scientists whose rigorous work is finally bearing fruit – not only in terms of benefits for patients, but also unprecedented insights into how psychedelics reset the brain. If the latest results stand up to closer scrutiny, they will transform the way we understand and treat mental illnesses.

The idea that they might be used to treat mental illness emerged in the 1950s, a decade or so after Swiss chemist Albert Hofmann first described his experiences of taking LSD. By the mid-1960s, roughly 40,000 people had been given LSD as part of treatments for all manner of mental illnesses, from obsessive compulsive disorder to addiction, depression and schizophrenia.

It looked like we were onto something. Then psychedelics escaped the lab and took off among the counterculture. The backlash meant that by 1970, they had been banned in the US, Canada and Europe. Research ground to a halt.

In the meantime, treatment for depression, the most common mental illness, came to be dominated by drugs called selective serotonin reuptake inhibitors (SSRIs), which boost levels of the neurotransmitter serotonin in synapses by blocking its reabsorption by neurons. Their success in early trials fuelled the idea that depression is caused by a deficiency in serotonin. But recently, this idea has been called into question, as more and more studies suggest SSRIs aren't as effective as we thought.

That comes as no surprise to many psychiatrists. Despite their ubiquity – 8.5 per cent of people in the US take them – SSRIs work for just 1 in 5 people. Even when they do work, there are problems, not least that coming off the drugs brings severe side effects. The picture is no less grim for other mental illnesses: there is a chronic shortage of new treatments and precious few ideas about where fresh options might come from.

That's part of the reason why a psychedelic revival has always been so tantalising. The first push came in the late 1990s, driven primarily by a US non-profit called the Multidisciplinary Association for Psychedelic Studies (MAPS). After a few individuals were determined enough to go through the arduous process of getting approval to work with psychedelics, the US Food and Drug Administration (FDA) decided to treat psychedelics like other drugs, meaning researchers were not banned from working with them.

Two decades later, those efforts are finally paying off. The psychedelic renaissance is entering a new stage, with a series of startling insights gracing the pages of leading journals and clinical trials making progress.

MDMA, better known as the party drug ecstasy, is the furthest along. Although not a classic psychedelic in that it doesn't induce hallucinations, MDMA works by flooding the brain with serotonin, which makes users feel euphoric. These mood-altering effects are the reason researchers became interested in using it as a tool to assist psychotherapy for people with post-traumatic stress disorder (PTSD).

PTSD will affect roughly 7 per cent of people in the US at some point in their lives. The most effective treatment involves memory reconsolidation. People are asked to recall traumatic events so that their memories of them can be stripped of fearful associations by processing them in a new way. The problem is that recall can sometimes be so terrifying that they have to stop receiving this form of therapy. MDMA appears to help, not only because it extinguishes anxiety and stress, but also because it triggers the release of oxytocin, a pro-social hormone that strengthens feelings of trust towards therapists.

At the Psychedelic Science 2017 conference in Oakland, California, a group led by Michael Mithoefer at the Medical University of South Carolina presented results from trials in which 107 people with PTSD underwent a psychotherapy while under the influence of MDMA. A year or so after having the therapy, roughly 67 per cent of them no longer had PTSD, according to a measure based on symptoms such as anxiety levels and frequency of nightmares. About 23 per cent of the control group, which had psychotherapy and a placebo drug, got the same benefit.

## Healing trip

That convinced the FDA to give the nod for Mithoefer's group to carry out further trials involving more participants, the last hurdle to clear before the drug can be approved. In fact, the FDA was so impressed that it granted MDMA 'break-through therapy' status, which will accelerate the path towards approval. If all goes well, it could be in use as soon as 2021.

If recent results are anything to go by, however, true psyche-delics – those that induce hallucinations – might end up having the biggest impact on mental health. That's because psilocybin, the active ingredient in magic mushrooms, is beginning to look like the real deal: a genuinely effective, long-lasting treatment for depression.

It started in 2006, when Roland Griffiths, a psychiatrist and neuroscientist at Johns Hopkins University in Baltimore, replicated the results of a notorious study from 1962. He showed that a large dose of psilocybin can induce mystical experiences in volunteers without any mental health prob-lems, including feelings of ego dissolution, a sense of revela-tion, ineffability and transcendence of time and space.

Fourteen months after taking the drug at Griffiths's lab, 22 of the 36 participants said the experience improved their well-being or life satisfaction, and rated it as one of the top five most meaningful experiences of their lives.

It was a landmark study. As Solomon Snyder, also at Johns Hopkins, wrote at the time: 'The ability of these researchers to conduct a double-blind, well-controlled study tells us that clinical research with psychedelic drugs need not be so risky as to be off-limits to most investigators.'

In a double-blind study, neither the researchers nor the participants know who is receiving the experimental treatment. It is tricky to do with drugs like psilocybin because the hallucinations they induce mean volunteers know they aren't taking a placebo. But Griffiths and his colleagues got around the problem by using a placebo that induces a slight stimulating effect to trick recipients into thinking they got the active drug.

Figuring that psychedelic experiences would be particularly valuable to people confronting a terminal illness, Griffiths and others began trials designed to assess the safety and efficacy of psilocybin to treat anxiety in people with advanced cancer. In the largest of those, Griffiths recruited 51 volunteers. Half of them were given a small placebo-like dose during one session, then a high dose five weeks later. For the other half, the sequence was reversed.

The results were published in 2016. There was a marked reduction in depression and anxiety symptoms compared with placebo after the high-dose session, and for 80 per cent of them those benefits continued to be felt six months later. An associated study at New York University reported similar results.

Meanwhile, Robin Carhart-Harris, a neuroscientist at Imperial College London, has been working with people with depression that has resisted all available treatments. In a trial involving 20 people, participants had two sessions – one on a single low dose of psilocybin (10 mg), one on a single high dose (25 mg) – during which they each separately lay listening to specially chosen music, accompanied by therapists.

The findings, also reported in 2016, were impressive. Those two doses, combined with the psychological support, were sufficient to lift depression in all 20 participants for three weeks, and to keep it at bay for five of them for three months.

That is in stark contrast to the best available antidepressants. 'What's weird and so different about these [psychedelics] is that we're talking about a single dose having long-term effects,' says Insel, now at a start-up called Mindstrong. 'That's a remarkably different approach to what we've been doing, with drugs that people take chronically.'

Hints as to why psychedelics work so quickly and so enduringly have come from brain scans. Since 2010, Carhart-Harris has used functional magnetic resonance imaging (fMRI) to scan the brains of people without mental illness while they are experiencing the effects of different psychedelic drugs. He has found that LSD and psilocybin both cause activity in parts of the brain that normally work separately to become more synchronous, meaning the neurons fire at the same time. In addition, connectivity across a collection of brain regions called the 'default mode network', which is linked to our sense of self, or ego, is drastically reduced. The more this network disintegrates, the more volunteers report a dissolving of the boundaries between themselves and the world around them.

Carhart-Harris thinks psilocybin therapy interrupts the spirals of rumination and negative thoughts that depressed people get caught up in. In that sense, it seemed telling that people in his psilocybin-for-depression trial who experienced aspects of a spiritual or mystical experience saw a bigger decrease in their depression scores than those who didn't.

To see what effect the drug had, however, Carhart-Harris and his colleagues scanned the brains of their participants before and after they received psilocybin-assisted therapy. Contrary to expectations, the integrity of the default mode network, meaning the extent to which neurons across its separate brain regions fire together, had increased one day after therapy. What's more, the magnitude of this effect correlated with the extent to which the volunteers' depression had lifted.

Since the volunteers weren't scanned during the acute drug experience, interpreting this result requires a bit of speculation, but Carhart-Harris sees this as a 'reset process'. 'You take something that's ordered, but pathologically ordered perhaps; you shock it and scramble it and then it returns, but it returns to a healthier mode,' he says.

For Carhart-Harris, this trick of unlocking the brain's ability to remodel itself, known as plasticity, is what makes psychedelics so unique and valuable. The effect isn't intrinsically therapeutic, he says, but when combined with psychotherapy it appears to have an unparalleled capacity to alleviate mental illness or behavioural problems.

## Back to the future?

The insights gleaned by peering into the brains of the people who volunteered for his psilocybin trial don't end there. Participants were shown pictures of happy and frightened faces as they lay in the fMRI machine. The amygdala, a part of the brain that deals with emotions, including fear, typically lights up in response to such stimuli. SSRIs dampen those responses. But after the combined psilocybin-psychotherapy session, the amygdala lit up. And again, this effect correlated with how well people did: the greater the response in the amygdala, the more their symptoms improved.

This suggests a profound change in the processing of emotions, which fits with what participants reported in interviews. While SSRIs blunt both positive and negative feelings, it seems psilocybin does the opposite, helping people reconnect with their emotions. Those may not always be positive, but the idea is that connection with emotions is better than numbness.

The usual caveats apply, of course: all of these studies are relatively small and Carhart-Harris's recent trial lacked a control group to directly contrast with those taking psilocybin. 'One needs to be cautious,' says Paul Summergrad at Tufts University in Boston, who is a former president of the American Psychiatric Association. 'The history of psychiatry and medicine is full of things people get excited about that don't play out.'

If larger studies produce similarly compelling outcomes, however, the implications would be profound. 'The conversation now with psilocybin and MDMA is very different than what we've had with the development of other antidepressants and anti-anxiety drugs,' says Insel. 'We're now talking about psilocybin-assisted therapy, meaning that it's not just about the chemical but the role the chemical can have in a psychotherapeutic experience,' he says.

For Insel, the fact that they are psychedelics is irrelevant. 'I'm excited to think that there might be compounds that could be used in a new way to give us something that will make a difference for people who haven't received much assistance from the drugs we have.'

So what now? The short answer is more trials. UK firm Compass Pathways plans to conduct a placebo-controlled psilocybin trial in 400 people with depression across eight European countries. Griffiths is also preparing for a placebo-controlled trial, and Carhart-Harris is planning one to compare psilocybin with a leading SSRI.

One problem is that drug development is an eye-wateringly expensive business. In preparation for MDMA being licensed for PTSD, however, MAPS has set up a public benefit corporation that will market the drug and use the profits to push through other promising psychedelics.

The biggest danger now might be that history repeats itself. The first wave of psychedelics research was to a great extent doomed by excessive enthusiasm. Today, as the revival has gathered steam, some doctors have likewise grown impatient and gone rogue, offering their patients underground psychedelic treatments. Hence the current crop of researchers are at pains to preach patience and rigour.

Insel put it more bluntly at the 2017 Psychedelic Science conference: 'Don't screw this up.'

## Fast asleep? Your unconscious is still listening

*Can you really set a mental alarm clock by hitting your head on the pillow before you go to bed? That's not so far from the truth, says **Simon Makin**.*

Some people swear that if they want to wake up at 6 a.m., they just bang their head on the pillow six times before going to sleep. Crazy? Maybe not. A study from 1999 shows that it all comes down to some nifty unconscious processing.

For three nights, a team at the University of Lübeck in Germany put 15 volunteers to bed at midnight. The team either told the participants they would wake them at 9 a.m. and did, or told them they would wake them at 9 a.m., but actually woke them at 6 a.m., or said they would wake them at 6 a.m. and did.

This last group had a measurable rise in the stress hormone adrenocorticotropin from 4.30 a.m., peaking around 6 a.m.. People woken unexpectedly at 6 a.m. had no such spike. The unconscious mind, the researchers concluded, can not only keep track of time while we sleep but also set a biological alarm to jump-start the waking process. The pillow ritual might help set that alarm.

The sleeping brain can also process language. In a 2014 study, Sid Kouider of the École Normale Supérieure in Paris and his colleagues trained volunteers to push a button with their left or right hand to indicate whether they heard the name of an animal or object as they fell asleep. The team monitored the brain's electrical activity during training and when the people heard the same words when asleep. Even when asleep, activity continued in the brain's motor regions, indicating that the sleepers were preparing to push the correct button. The people could also correctly categorise new words, first heard after they had dropped off, showing that they were genuinely analysing the meaning of the words while asleep.

It's an ability that makes good evolutionary sense, says Kouider. 'If you stop monitoring your environment, you become very vulnerable during sleep . . . It makes sense that you don't simply shut down, but continue tracking in a kind of standby mode.' This might explain why some sounds, like our names, wake us more easily than others.

This protective monitoring may not last all night, however. A study published in 2016 found that while language processing continues in REM sleep for words heard just before bed, once in deep sleep all responses disappear as the brain goes 'offline' to allow the day's memories to be processed. 'Your cognition about things in the environment declines progressively towards deep sleep,' Kouider says. 'Sleep is not all-or-none in terms of cognition, it's all-or-none in terms of consciousness.'

## How your brain works things out all by itself

*You may have taken a break, but your brain hasn't.*
**Caroline Williams** *reveals how the unconscious carries on mulling things over long after you quit.*

Wouldn't it be great if you could leave difficult decisions to your subconscious, secure in the knowledge that it would do a better job than conscious deliberation? Ap Dijksterhuis of Radboud University Nijmegen in the Netherlands proposed this counter-intuitive idea in 2004. No wonder it was instantly popular.

Dijksterhuis had found that volunteers asked to make a complex decision – such as choosing between different apartments based on a baffling array of specifications – made better choices after being distracted from the problem before deciding. He reasoned that this is because unconscious thought can move beyond the limited capacity of working memory, so it can process more information at once.

The idea has been influential, but it may be too good to be true. Many subsequent studies have failed to replicate Dijksterhuis's results. And a recent analysis concluded that there is little reason to think the unconscious is the best tool for making complex decisions. Still, Dijksterhuis remains confident that the effect is real and is an important part of our mental toolkit.

Others think the unconscious mind's way of processing

information is more important for creativity than for decision-making. It brings together disparate information from all over the brain without interference from the brain's goal-directed frontal lobes. This allows it to generate novel ideas that burst through to consciousness in a moment of insight. John Kounios of Drexel University in Philadelphia believes an idea can only be truly creative if it appears in this way.

Some people seem to be better wired for this kind of thinking. Kounios has found that people who tend to solve problems in 'aha' moments of insight have different resting state brain activity – with less frontal control – than more logical thinkers.

While there is no known way to change your brain into a more creative one, Kounios suggests thinking about a problem until you get stuck, then taking a break and hoping that something useful bubbles up before your deadline.

## How we weigh up a person's character in 0.1 seconds

*Your unconscious mind is seriously judgemental. But our snap decisions often turn out to be spot on.* **Simon Makin** *explains how.*

Ever felt love at first sight? Or an irrational distrust of a stranger on a bus? It could be because our unconscious is constantly making fast judgements. And they are often pretty accurate.

In the early 1990s, Nalini Ambady and Robert Rosenthal, both then at Stanford University in California, asked volunteers to rate teachers on traits including competence, confidence and honesty after watching 2-, 5- or 10-second silent clips of their performance. The scores successfully predicted the teachers' end of semester evaluations and 2-second judgements were as accurate as those given more time. Further experiments showed similar accuracy for judgements about sexuality, economic success and political affiliation. For anyone hoping to use this to their advantage, the bad news is that no one has worked out what to do to pass yourself off as a winner. It seems to be an overall body signal that is both given out and picked up unconsciously, and is greater than the sum of its parts. This makes it very difficult if not impossible to fake.

In some cases, all we need to make these judgements is a glimpse of a face. In a separate study, people saw the faces of US election candidates for 1 second and were then asked to rate their competence – these ratings not only predicted the winning candidates, but also their margin of victory. A follow-up study found that people could make such judgements given only a tenth of a second. Again, the magic ingredients of what makes a face you can trust haven't been identified, so this is one area of the unconscious where we have little choice in the conclusions we draw. While the skill is undoubtedly useful, it can also make unfounded prejudices feel like intuition when they are actually the result of our unconsciously held biases towards specific social groups.

Although we can't easily change our facial features, our unconscious mind has a trick for making us likeable: mimicry. Jo Hale, a psychologist at University College London, is using virtual avatars to study the popular idea that we like people who mimic our body language. While it takes a lot of effort to consciously mimic someone's body language, we do it effortlessly, without thinking, all the time. In a recent study, Hale programmed virtual avatars to mimic volunteers with a 1- or 3-second delay in their mimicry and found that 3 seconds may be close to a natural delay, because it rendered people both unaware they were being mimicked and more likely to rate the avatar as likeable. A delay of 1 second seemed to raise a flag to the consciousness, making volunteers more likely to notice the mimicry. So despite what body language coaches might have you believe, mimicry may only work if you get the timing right.

## We know where our limbs are without thinking

*Your unconscious has a sixth sense of the space your body takes up, and the invisible area around it.* **Anil Ananthaswamy** *discovers that getting to know it better could improve your memory.*

Thanks to unconscious processing, most of us instinctively know where our limbs are and what they are doing. This ability, called proprioception, results from a constant conversation between the body and brain. This adds up to an unerring sense of a unified, physical 'me'.

This much-underrated ability is thought to be the result of the brain predicting the causes of the various sensory inputs it receives – from nerves and muscles inside the body, and from the senses detecting what's going on outside the body. 'What we become aware of is the brain's "best guess" of where the body ends and where the external environment begins,' says Arvid Guterstam of the Karolinska Institute in Stockholm.

The famous rubber-hand illusion is a good example of this. In this experiment, a volunteer puts one hand on the table in front of them. Their hand is hidden, and a rubber hand is put in front of the participant. A second person then strokes the real and rubber hands simultaneously with a paintbrush. Within minutes, many people start to feel the strokes on the rubber hand, and even claim it as part of their body. The brain is making its best guess as to where the sensation is coming from and the most obvious option is the rubber hand.

Recent research suggests this sixth sense extends to the space immediately surrounding the body. Guterstam and his colleagues repeated the experiment, stroking the real hand but keeping the brush 30 centimetres above the rubber hand. Participants still sensed the brush strokes above the rubber hand, implying that as well as unconsciously monitoring our body, we keep track of an invisible 'force field' around us. Guterstam suggests this might have evolved to help us pick up objects and move through the environment without injury.

## Move to improve

A lack of proprioception is rare but can happen with nerve or brain damage. The case of Ian Waterman, who lost proprioception after nerve damage caused by a flu-like virus in 1971, demonstrates just how much we rely on this ability. After being told he would never walk again, he slowly learned to consciously control his muscles to move his body. Decades later, it is still far from easy and he only has full control over his movements if he is looking at the relevant body part and concentrating. 'Because his proprioceptive system is shot, these things are not automatic for him. It requires constant conscious effort,' says Anil Seth, a neuroscientist at the University of Sussex in Brighton.

Even if the system is working fine, there is some evidence that it might be worth consciously trying to improve it. A recent study in which volunteers trained in MovNat exercise – a programme designed to tax the body's natural balancing, jumping and vaulting abilities – improved more on measures of working memory than a control group who did yoga or no exercise.

## Your brain's crystal ball helps you understand speech and fear

*Matching what your brain predicts to what actually happens gives you a jump-start on how to react, writes* **Diana Kwon**, *but what happens when your expectations go awry?*

Every moment, the brain takes in far more information than it can process on the fly. In order to make sense of it all, the brain constantly makes predictions that it tests by comparing incoming data against stored information. All without us noticing a thing.

Simply imagining the future is enough to set the brain in motion. Imaging studies have shown that when people expect a sound abstract or image to appear, the brain generates an anticipatory signal in the sensory cortices.

This ability to be one step ahead of the senses has an important role in helping us understand speech. 'The brain is continuously predicting the sounds, words and meanings that people are trying to produce or communicate,' says Matt Davis at the MRC Cognition and Brain Sciences Unit in Cambridge.

Studies have also shown that the brain can use one sense to inform another. When you hear a recording of speech that is so degraded it is nearly unintelligible, the words sound clearer if you have previously read the same words in subtitles. 'The sensory parts of the brain are comparing the speech you've heard to the speech you predicted,' says Davis.

Not only do we make hypotheses about external information,

our brains also make predictions on the basis of emotional signals coming from the body. Moshe Bar, a neuroscientist at Bar-Ilan University in Israel, goes so far as to suggest that we consciously recognise an object only once our unconscious mind has calculated its importance based on what our senses and emotional reaction are saying. The conscious fear of a snake on a hiking trail comes after the brain has processed the shape and initiated jumping out of the way, for example.

Making predictions does have its downsides, however. Incorrect inferences reinforced by repetition can be hard to reverse, which is why when you learn the wrong lyrics to a song, it can be difficult to stop hearing them. Stereotyping is a more troublesome example of the same thing. While it can be useful to recognise that the dangers of things like snakes and fires are relatively constant, when it comes to human interactions, it can lead to negative biases and discrimination. 'Stereotypes and prejudices are predictions working as they do with everything else, but [in a way] that is not desirable,' says Bar.

Some neuroscientists also believe that the hallucinations experienced in psychosis are the result of expectations gone awry. In one recent study, people who were more prone to psychotic experiences were better at seeing hidden shapes in images that had been digitally degraded. The researchers speculate that this could mean their brains jump to conclusions faster and rely less on evidence coming in from the senses.

Despite its flaws, prediction is hugely beneficial. 'Imagine that our brain didn't work like that,' says Bar. 'Every snake you see you'd have to learn afresh. Every fire you'd have to touch and burn yourself.'

## You can break bad habits by hacking the autopilot in your brain

*Do something enough times and your brain can automate the process, making good habits and bad.* **Anil Ananthaswamy** *reveals the ways you can get back some conscious control.*

So much of what we do in our day-to-day lives, whether it be driving, making coffee or touch-typing, happens without the need for conscious thought. Unlike many of the brain's other unconscious talents, these are skills that have had to be learned before the brain can automate them. How it does this might provide a method for us to think our way out of bad habits.

Ann Graybiel of the Massachusetts Institute of Technology and her colleagues have shown that a region deep inside the brain called the striatum is key to habit forming. When you undertake an action, the prefrontal cortex, which is involved in planning complex tasks, communicates with the striatum, which sends the necessary signals to enact the movement. Over time, input from the prefrontal circuits fades, to be replaced by loops linking the striatum to the sensorimotor cortex. The loops, together with the memory circuits, allow us to carry out the behaviour without having to think about it. Or, to put it another way, practise makes perfect. No thinking required.

The upside of this two-part system is that once we no longer need to focus our attention on a frequent task, the spare processing power can be used for other things. It comes with a downside, however. Similar circuitry is involved in turning all kinds of behaviours into habits, including thought patterns, and once any kind of behaviour becomes habit, it becomes less flexible and harder to interrupt. 'If it's a good habit, that's absolutely fine,' says neuroscientist Anil Seth at the University of Sussex. 'But if you ingrain a bad habit, that's equally difficult to get rid of. You lose that moment of choice when you can decide not to do something.'

Crucially, though, Graybiel's team has shown that, even with the most ingrained habits, a small area of the prefrontal cortex is kept online, in case we need to take alternative action. If the brake pedal in our car stops working, for instance, our entire focus of attention shifts to the physical act of driving the car. This offers hope to anyone looking to break a bad habit, and to those suffering from habit-related problems such as obsessive–compulsive disorder and Tourette's syndrome – both of which are associated with abnormal activity in the striatum and its connections to other parts of the brain. These circuits could prove fruitful targets for future drug treatments. For now, though, the best way to get a handle on bad habits is to become aware of them. Then, focus all your attention on them and hope that it's enough to help the frontal regions resist the call of the autopilot. Or you could teach yourself a new habit that counters the bad one.

## How you hallucinate to make sense of the world

*Understanding what is happening in the brain during hallucinations reveals how we're having them all the time, and how they shape our perception of reality, says **Helen Thomson**.*

Avinash Aujayeb was alone, trekking across a vast white glacier in the Karakoram, a mountain range on the edge of the Himalayan plateau known as the roof of the world.

Although he had been walking for hours, his silent surroundings gave little hint that he was making progress. Then suddenly, his world was atilt. A massive icy boulder loomed close one moment, but was desperately far away the next. As the world continued to pulse around him, he began to wonder if he could believe his eyes. He wasn't entirely sure he was still alive.

A doctor, Aujayeb checked his vitals. Everything seemed fine: he wasn't dehydrated, nor did he have altitude sickness. Yet the icy expanse continued to warp and shift. Until he came upon a companion, he couldn't shake the notion that he was dead.

In recent years it has become clear that hallucinations are much more than a rare symptom of mental illness or the result of mind-altering drugs. Their appearance in those of sound mind has led to a better understanding of how the brain can create a world that doesn't really exist. More surprising, perhaps, is the role they may play in our perceptions of the real world. As researchers explore what is happening in the brain, they are beginning to wonder: do hallucinations make up the very fabric of our reality?

Hallucinations are sensations that appear real but are not elicited by anything in our external environment. They aren't only visual – they can be sounds, smells, even experiences of touch. It's difficult to imagine just how real they seem unless you've experienced one. As Sylvia, a woman who has had musical hallucinations for years, explains, it's not like imagining a tune in your head – more like 'listening to the radio'.

There is evidence to support the sensation that these experiences are authentic. In 1998, researchers at King's College London scanned the brains of people having visual hallucinations. They found that brain areas that were active are also active while viewing a real version of the hallucinated image.

Those who hallucinated faces, for example, activated areas of the fusiform gyrus, known to contain specialised cells active when we look at real faces. The same was true with hallucinations of colour and written words. It was the first objective evidence that hallucinations are less like imagination and more like real perception.

Their convincing nature helps explain why hallucinations have been given such meaning – even considered messages from gods. But as it became clear that they can be symptoms of mental illnesses such as schizophrenia, they were viewed with increasing suspicion.

We now know that hallucinations occur in people with perfectly sound mental health. The likelihood of experiencing them increases in your sixties; 5 per cent of us will experience one or more hallucinations in our life.

Many people hallucinate sounds or shapes before they drift off to sleep, or just on waking. People experiencing extreme grief have also been known to hallucinate in the weeks after their loss – often visions of their loved one. But the hallucinations that may reveal the most about how our brain works are those that crop up in people who have recently lost a sense.

I have personal experience of this. At 87, my grandmother began to hallucinate after her already poor sight got worse due to cataracts. Her first visitors were women in Victorian dress, then young children. She was experiencing what is known as Charles Bonnet syndrome. Bonnet, a Swiss scientist who lived in the early 1700s, first described the condition in his grandfather, who had begun to lose his vision. One day the older man was sitting talking to his granddaughters when two men appeared, wearing majestic cloaks of red and grey. When he asked why no one had told him they would be coming, the elder Bonnet discovered only he could see them.

It's a similar story with Sylvia. After an ear infection

caused severe hearing loss, she began to hallucinate a sound that was like a cross between a wooden flute and a bell. At first it was a couple of notes that repeated over and over. Later, there were whole tunes. 'You'd expect to hear a sound that you recognise, maybe a piano or a trumpet, but it's not like anything I know in real life,' she says.

Max Livesey was in his seventies when Parkinson's disease destroyed the nerves that send information from the nose to the brain. Despite his olfactory loss, one day he suddenly noticed the smell of burning leaves. The odours intensified over time, ranging from burnt wood to a horrible onion-like stench. 'When they're at their most intense they can smell like excrement,' he says. They were so powerful they made his eyes water.

Sensory loss doesn't have to be permanent to bring on such hallucinations. Aujayeb was in fine health, trekking across the glacier. 'I felt very tall – the ground appeared far from my eyes. It was like I was seeing the world from over my shoulder,' he explained. His hallucinations continued for 9 hours, but after a good night's sleep, they were gone.

When our senses are diminished, all of us have the potential to hallucinate. It can take just 30 to 45 minutes for people to experience hallucinations if they try a simple visual deprivation technique. In a study run by Jiří Wackermann at the Institute for Frontier Areas of Psychology and Mental Health in Freiburg, one volunteer saw a jumping horse. Another saw an eerily detailed mannequin. 'It was all in black . . . had a long narrow head, fairly broad shoulders, very long arms.'

Yet why should a diminished sense trigger a sight, sound or smell that doesn't really exist? 'The brain doesn't seem to tolerate inactivity,' said the late neurologist Oliver Sacks when I spoke to him about this in 2014. 'The brain seems to respond to diminished sensory input by creating autonomous sensations of its own choosing.' This was noted soon after the Second World War, he said, when it was discovered that high-flying aviators in featureless skies and truck drivers on long, empty roads were prone to hallucinations.

Now researchers believe these unreal experiences provide a glimpse into the way our brains stitch together our perception of reality. Although bombarded by thousands of sensations every second, the brain rarely stops providing you with a steady stream of consciousness. When you blink, your world doesn't disappear. Nor do you notice the hum of traffic outside or the tightness of your socks. Well, you didn't until they were mentioned. Processing all of those things all the time would be a very inefficient way to run a brain. Instead, it takes a few shortcuts.

Let's use sound as an example. Sound waves enter the ear and are transmitted to the brain's primary auditory cortex, which processes the rawest elements, such as patterns and pitch. From here, signals get passed on to higher brain regions that process more complex features, such as melody and key changes.

Instead of relaying every detail up the chain, the brain combines the noisy signals coming in with prior experiences to generate a prediction of what's happening. If you hear the opening notes of a familiar tune, you expect the rest of the song to follow. That prediction passes back to lower regions, where it is compared to the actual input, and to the frontal lobes, which perform a kind of reality check, before it pops up into our consciousness. Only if a prediction is wrong does a signal get passed back to higher areas, which adjust subsequent predictions.

This idea is consistent with what was happening to Sylvia. Although she was mostly deaf, she could still make out some sound – and she discovered that listening to familiar Bach concertos suppressed her hallucinations. Timothy Griffiths, a cognitive neurologist at Newcastle University, scanned Sylvia's brain before, after and while listening to Bach, and had her rate the intensity of her hallucinations throughout. They were at their quietest just after the real music was played, gradually increasing in volume until the next excerpt.

The brain scans showed that during her hallucinations, the higher regions that process melodies and sequences of tones were talking to one another. Yet, because Sylvia is severely deaf, they were not constrained by the real sounds entering her ears. Her hallucinations are her brain's best guess at what is out there.

The notion of hallucinations as errant predictions has also been put to the test in completely silent rooms known as anechoic chambers. The quietest place on earth is one such chamber at Orfield Laboratories in Minneapolis, Minnesota. Once inside, you can hear your eyeballs moving. People generally start to hallucinate within 20 minutes of the door closing. But what's the trigger?

There are two possibilities. One is that sensory regions of the brain sometimes show spontaneous activity that is usually suppressed and corrected by real sensory data coming in from the world. In the deathly silence of an anechoic chamber, the brain may make predictions based on this spontaneous activity. The second possibility is that the brain misinterprets internally generated sounds, says Oliver Mason at University College London. The sound of blood flowing through your ears isn't familiar, so it could be misattributed as coming from outside you. 'Once a sound is given significance, you've got a seed,' says Mason, 'a starting point on which a hallucination can be built.'

Not everyone reacts the same way inside an anechoic chamber. Some people don't hallucinate at all. Others do, but realise it was their mind playing tricks. 'Some people come out and say "I'm convinced you were playing noises in there",' says Mason.

Understanding why people react differently to a diminished sensory environment could reveal why some are more prone to the delusions and hallucinations associated with mental illness. We know that electrical messages passed across the brain are either excitatory or inhibitory – meaning they either promote or impede activity in neighbouring neurons. In recent experiments, Mason's team scanned the brains of volunteers as they sat in an anechoic chamber for 25 minutes. Those who had more hallucinatory sensations had lower levels of inhibitory activity across their brain. Perhaps, says Mason, weaker inhibition makes it more likely that irrelevant signals suddenly become meaningful.

People with schizophrenia often have overactivity in their sensory cortices, but poor connectivity from these areas to their frontal lobes. So the brain makes lots of predictions that are not given a reality check before they pass into conscious awareness, says Flavie Waters, a clinical neuroscientist at the University of Western Australia in Perth. In conditions like Charles Bonnet syndrome, it is underactivity in the sensory cortices that triggers the brain to start filling in the gaps, and there is no actual sensory input to help it correct course. In both cases, says Waters, the brain starts listening in on itself, instead of tuning into the outside world. Something similar seems to be true of hallucinations associated with some recreational drug use.

As these insights help us to solve the puzzle of perception, they are also providing strategies for treating hallucinations. People with drug-resistant schizophrenia can sometimes reduce their hallucinatory symptoms by learning how to monitor their thoughts, understand the triggers and reframe their hallucinations so that they see them in a more positive and less distressing light. 'You're increasing their insight and their ability to follow their thoughts through to more logical conclusions,' says Waters. This seems to give them more control over the influence of their internal world.

This kind of research is also helping people like Livesey reconnect with the external world. If his phantosmia, or smell hallucinations, are driven by a lack of reliable information, then real smells should help him to suppress the hallucinations. He has been trialling sniffing three different scents, three times a day. 'Maybe it's just wishful thinking,' he says, 'but it seems to be helping.'

The knowledge that hallucinations can be a by-product of how we construct reality might change how we experience them. In his later years, Sacks experienced hallucinations after his eyesight began to fail. When he played the piano, he would occasionally see showers of flat symbols when he was looking carefully at musical scores. 'I have long since learned to ignore my hallucinations, and occasionally enjoy them,' said Sacks. 'I like seeing what my brain is up to when it is at play.'

## A hole in the head could help stall dementia

*Our ancestors used to drill holes in the skull to expel demons – and the technique has made a comeback as a cure for dementia.* **Arran Frood** *drills down.*

In the early 1960s, a young Russian neurophysiologist called Yuri Moskalenko travelled from the Soviet Union to the UK on a Royal Society exchange programme. During his stay, he co-authored a paper published in *Nature*. 'Variation in blood volume and oxygen availability in the human brain' may not sound subversive, but it was the start of a radical idea.

Decades later, having worked in Soviet Russia and become president of the Sechenov Institute of Evolutionary Physiology and Biochemistry at the Russian Academy of Sciences in St Petersburg, Moskalenko returned to the UK. He began collaborating with researchers at the Beckley Foundation in Oxford, and in 2010 his work started to bear fruit.

And weird fruit it is. With funding from the foundation, he is exploring the idea that people with Alzheimer's disease could be treated by drilling a hole in their skull. In fact, he is so convinced of the benefits of trepanation that he claims it may help anyone from their mid-forties onwards to slow or even reverse the process of age-related cognitive decline. Can he be serious?

For thousands of years, trepanation has been performed for quasi-medical reasons such as releasing evil spirits that were believed to cause schizophrenia or migraine. Today it is used to prevent brain injury by relieving intracranial pressure, particularly after accidents involving head trauma.

In the popular imagination, though, it is considered crude, if not downright barbaric. Yet such is the desperation for effective treatments for dementia that drilling a hole in the skull is not even the strangest game in town.

The problem is huge and growing. Alzheimer's, the most common form of dementia, affects 700,000 people in the UK and nearly 5 million in the US. In addition, 1 in 5 Americans over the age of 75 have mild cognitive impairment, which often leads to Alzheimer's. As people live longer, the numbers seem certain to grow. Yet we have few ideas about what causes dementia and even fewer about how to treat it. Most patients get a mixture of drugs and occupational therapy, which at best stalls the apparent progression of their illness by masking its symptoms.

The causes of dementia are many and poorly understood,

but there is growing evidence that one factor is the flow of blood within the brain. As we age, cerebral blood flow decreases, and the earlier this happens the more likely someone is to develop early onset dementia. It remains unclear, however, whether declining cerebral blood flow is the cause, or an incidental effect of a more fundamental change. Moskalenko's research indicates that cerebral blood flow is more closely correlated with age than with levels of dementia, so he decided to delve more deeply.

## The brain's buffer

As well as delivering oxygen to the brain, cerebral blood flow has another vital role: the circulation and production of cerebrospinal fluid. This clear liquid surrounds the brain, carrying the nutrients that feed it and removing the waste it produces, including the tau and beta-amyloid proteins that have been implicated in the formation of plaques found in the brains of people with Alzheimer's.

How blood flow influences cerebrospinal fluid flow can be gauged from something called 'cranial compliance', a measure of the elasticity of the brain's vascular system. 'The cranium is a bony cavity of fixed volume, with the brain taking up most of the space,' says Robin Kennett, a neurophysiologist from the Oxford Radcliffe Hospitals in the UK. 'Every time the heart beats and sends blood into the cranium, something else has to come out to prevent the pressure rising to levels that would damage the brain.' So, as fresh blood flows into the brain's blood vessels, cerebrospinal fluid flows out into the space around the spinal cord through a hole in the base of the skull called the foramen magnum.

As we age, the proteins in the brain harden, preventing this system from working as it should. As a result, the flow of both blood and cerebrospinal fluid is reduced, impairing the delivery of oxygen and nutrients as well as the removal of waste. Moskalenko's research suggests that this normally begins between the ages of 40 and 50. Moreover, in a study of 42 elderly people with dementia, he found that the severity of their cognitive disorder was strongly correlated with cranial compliance: those with the severest dementia had the lowest compliance. 'Cranial compliance is a significant component of the origin of certain cases of brain pathology,' he says.

This view gets qualified agreement from Conrad Johanson, a clinical neuroscientist at Brown University in Providence, Rhode Island. Although the link between low compliance and dementia has yet to be comprehensively shown, he says, 'there's a gestalt that it's broadly true'.

So where does trepanation come into all this? 'A hole made in the bony cavity would act as a pressure-release valve,' says Kennett, and this would alter the flow of fluids around the brain. This is exactly what Moskalenko observed when he carried out one of the first neurophysiological studies on trepanation.

Moskalenko studied 15 people who had undergone the procedure following head injuries. He found that their cranial compliance was around 20 per cent higher than the average for their age. Based on this, he calculates that a 4 cm$^2$ hole increases cerebral blood flow by between 8 and 10 per cent, which is equivalent to 0.8 millilitres more blood per heartbeat. This, he says, shows that trepanation could be an effective treatment for Alzheimer's, and he even goes so far as to suggest that it might provide a 'significant' improvement in the mental functions of anyone from their mid-forties, when cranial compliance starts to decline.

## Spinal taps

Surprisingly, his most vociferous critics do not challenge his support for trepanation. Instead they question his ideas about how it works. Gerald Silverberg at the Stanford School of Medicine in California points out that drilling a hole in the skull may temporarily drain the cranial cavity of cerebrospinal fluid together with any toxins that may have accumulated in it, effectively flushing out the system. 'Metabolite clearance, or the lack of it, is felt to be an important factor in the development of age-related dementias,' he says. A similar intervention, known as a lumbar shunt or 'spinal tap', in which a needle is inserted into the spinal column to remove cerebrospinal fluid, can dramatically improve the cognitive performance of people who undergo the procedure, Silverberg says. Spinal taps are normally used as a treatment for hydrocephalus – water on the brain – but Silverberg is now trying it out on people with Alzheimer's, and early studies suggest it helps.

Olivier Baledent, a neurophysiologist based at the University Hospital of Amiens, France, also questions Moskalenko's focus on cranial compliance. Like Silverberg, he believes cerebrospinal fluid itself is key. Baledent's research shows that in people with mild cognitive impairment, there is reduced activity in a part of the brain called the choroid plexus, where cerebrospinal fluid is formed. He suspects this results in impaired fluid formation and reabsorption, leading to a build-up of toxins, and that a spinal tap may be able to stop or decrease dementia by improving fluid turnover. Trepanation could work in a similar way.

So will dementia patients and their families ever accept trepanation as a treatment for the condition? Johanson, who sees trepanation as no more alarming than a spinal tap, admits that it is always going to be a hard sell. 'People think it's witchcraft when you drill a hole in the skull and patients are improving.'

Harriet Millward, when deputy chief executive of UK-based charity Alzheimer's Research Trust, was keeping an open mind. 'The procedure has been understudied so far and, until further research has been undertaken, the possibility of beneficial effects remains open,' she said. David Smith, a neuropharmacologist and head of the Oxford Project to Investigate Memory and Ageing, is more receptive. 'I think the observations look pretty robust,' he says. In the absence of drug treatments for dementia, 'these rather drastic surgical ones are worth considering', he says.

## Meet the people who can see time

*For some the year is a C stretching out in front of them, for others it's a hula hoop.* **Caroline Williams** *discovers that the ways some people visualise calendars could shed light on memory itself.*

For Emma, the end of the year has special significance, and not just because of all the gifts and food. It's also the only time of year when the date in her mental calendar lines up perfectly with her body.

Emma is a calendar synaesthete, one of a handful of people who see time: not as a vague conceptual timeline, but as a vivid calendar that feels so real they could almost touch it. This is a little-known variation of synaesthesia, in which the brain links one kind of sensation to another. Some people associate shapes with certain sounds, or colours with numbers. Emma sees time as a hula hoop, which anchors 31 December to her chest and projects the rest of the year in a circle that extends about a metre in front of her.

Heidi, another calendar synaesthete, sees the year as a backwards C hovering before her, with January at one end of the horseshoe and December at the other. When she thinks of a date she feels herself travel along the calendar to the right spot. She has a separate, hoop-shaped calendar for days of the week. Both have been part of her life for as long as she can remember.

The fact that certain people can vividly conjure number lines and calendars was first noted by Victorian polymath Francis Galton in 1880, but we have only recently begun to figure out how – and why. It's not just a matter of idle curiosity. Understanding how calendar synaesthesia works may help unravel the way we all keep track of our memories as we move through space and time.

That's because calendar synaesthetes experience a super-charged version of the way everyone else experiences time. Studies of different cultures around the world have shown that our perceptions vary slightly – most people in the West perceive time as a straight line running through their bodies, with the future ahead of them, while in parts of Papua New Guinea time flows uphill and for some Chinese people it flows downwards. But we all compute the abstract concept of time in the same way: in our brains, 'time is always mapped onto space,' says V. S. Ramachandran, a neuroscientist at the University of California, San Diego.

The mapping job falls largely to the hippocampus, a pair of curved structures towards the centre of the brain that contain specialised neurons. Some, called grid cells, plot locations, while others, known as place cells, become active when we arrive on the scene. The basic circuitry seems to have evolved about 300 million years ago in a fish-like common ancestor, and similar systems are found in most other animals, from lizards to birds. At some point in human evolution, though, the hippocampus gained a second role:

storing autobiographical memories, each with a time stamp recorded by specialised time cells.

'As you live your life, place cells keep track of your location in the world, and time cells keep track of stimuli receding into the past,' says neuroscientist Marc Howard at Boston University. 'When you vividly remember a specific event from your life – say lunch last Tuesday – the hippocampus recovers the activity of time cells and place cells that were active during that event.'

Whether any other animals have this kind of autobiographical memory is hotly debated, but we know for sure that no other species makes calendars. Around 10,000 years ago, we began to notice the natural cycles of the sun and moon and record them for future reference, first in stone circles, and today on paper and computer screens.

But calendar synaesthetes don't need to. They can call up their mental versions at will, something most are surprised to learn is unusual. Heidi first realised in a psychology class in high school. 'My teacher was talking about synaesthesia and how some people see calendars. I said, "Doesn't everybody see a calendar? How can you not?"'

Ramachandran wanted to know how they do it, and if they were really seeing calendars or summoning something from memory. So he asked a 20-year-old synaesthete called ML to recite alternate months between January and December, first forwards and then backwards. For most people, it takes three times as long to go backwards, because we have to construct the calendar from memory as we go. But ML was equally fast in both directions. She also unconsciously moved her eyes and finger as she went, suggesting her calendar was always in front of her.

To find out more, Ramachandran also used visual illusions, including the 'motion after-effect'. If you stare for 30 seconds at a contracting spiral and then look at another picture, it will appear to expand, because the brain's prediction outpaces our perception. But the illusion doesn't happen if you look at a blank wall or just imagine a scene. 'The brain needs something to attribute it to,' says Ramachandran.

When ML looked at her calendar after the spiral, it expanded in the same way as a real image. When asked to imagine an object in her mind's eye, it stayed still. That means that, as far as her brain is concerned, the calendar isn't a figment of her imagination, it is actually there.

What is going on? Ramachandran points to an area of the brain that we rely on to make sense of symbols and numbers and order events into sequences. The angular gyrus is found above and behind the ears on each side of the brain at the junction of several sensory areas, including the visual cortex. It also connects directly to the hippocampus. We all probably use this bit of circuitry to imagine the layout of time, but Ramachandran believes this is where calendar synaesthetes have the extra connections that make their visions so very real.

There are many open questions, not least whether this vivid calendar helps memory. There's reason to think so. 'If you ask them about a specific memory, then they'll conjure up the calendar and put the memory in the appropriate slot,' says Ramachandran.

That might be a trick worth learning. Daniel Bor at the University of Sussex has found that people can teach themselves to experience synaesthesia by repeatedly associating colours with certain letters. It might be possible to do something similar with calendars.

But they may not be a universal boon. One synaesthete Ramachandran met finds her calendar confusing, and another says hers is missing August, which can be frustrating – not least for making plans for a summer break.

For Heidi, it's a mixed bag. 'It helps me sometimes because I can picture things better, but I do get mixed up.' Her horseshoe-shaped calendar has a big gap after December, which means January always comes sooner than she expects it to. 'It feels really abrupt, like a whole month was in between them and it just went all of a sudden,' she says. Returning to the office after Christmas, that's probably something we can all relate to.

## The awesome emotion that gives us superpowers

*Awe is so powerful it alters your sense of self, connects you with humanity and boosts your mind and body, writes* **Jo Marchant**. *And there's a surprising way to get more of it.*

Have you ever been stopped in your tracks by a stunning view, or gobsmacked by the vastness of the night sky? Have you been transported by soaring music, a grand scientific theory or a charismatic person? If so, you will understand US novelist John Steinbeck's response to California's giant redwood trees, which can soar more than a hundred metres towards the sky. '[They] leave a mark or create a vision that stays with you always,' he wrote. 'From them comes silence and awe.'

Philosophers and writers have long been fascinated by our response to the sublime, but until a few years ago, scientists had barely studied it. Now they are fast realising that Steinbeck was right about its profound effects. Feeling awestruck can dissolve our very sense of self, bringing a host of benefits, from lowering stress and boosting creativity to making us nicer people.

Yet in the modern world, the value of the word awesome has plummeted – almost anything can now acquire the epithet. At the same time, we risk losing touch with the most potent sources of awe. The good news is that there are ways to inject more of it into our everyday lives. You needn't be religious. All you need is an open mind – although a willingness to try psychedelic drugs may help.

But what exactly is awe and where does it come from? 'It's a subjective feeling rooted in the body,' according to psychologist and pioneering awe researcher Dacher Keltner at the University of California, Berkeley. In 2003, he and Jonathan Haidt, now at New York University, published the first scientific definition. They described awe as the feeling we get when confronted with something vast, that transcends our frame of reference and that we struggle to understand. It's an emotion that combines amazement with an edge of fear. Wonder, by contrast, is more intellectual – a cognitive state in which you are trying to understand the mysterious.

You might think that investigating such a profound experience would be a challenge, but Keltner insists it's not so hard. 'We can reliably produce awe,' he says. 'You can get people to go out to a beautiful scene in nature, or put them in a cathedral or in front of a dinosaur skeleton, and they're going to be pretty amazed.' Then, all you need is a numerical scale on which people can report how much awe they are feeling. Increasingly, studies are including a physiological measure too, such as the appearance of goosebumps – awe is the emotion most likely to cause them, and second only to cold as a source.

In this way, Keltner and others have found that even mild awe can change our attitudes and behaviour. For example, people who watched a nature video that elicited awe – rather than other positive emotions such as happiness or pride –

were subsequently more ethical, more generous and described themselves as feeling more connected to people in general. Gazing up at tall eucalyptus trees left others more likely to help someone who stumbled in front of them. And after standing in front of a *Tyrannosaurus rex* skeleton, people were more likely to describe themselves as part of a group. It might seem counter-intuitive that an emotion we often experience alone increases our focus on others. But Keltner thinks it's because awe expands our attention to encompass a bigger picture, so reduces our sense of self.

'The desert is so huge, and the horizons so distant, that they make a person feel small,' wrote Paulo Coelho in *The Alchemist*. He was right. In a large study, Keltner found that after inspiring awe in people from the US and China, they signed their names smaller and drew themselves smaller, but with no drop in their sense of status or self-esteem. Similarly, neuroscientist Michiel van Elk at the University of Amsterdam found that people who watched awe-inducing videos estimated their bodies to be physically smaller than those who watched funny or neutral videos.

The cause of this effect might lie in the brain. In June 2017, at the annual meeting of the Organization for Human Brain Mapping in Vancouver, van Elk presented functional MRI scans showing that awe quiets activity in the default mode network, which includes parts of the frontal lobes and cortex, and is thought to relate to the sense of self. 'Awe produces a vanishing self,' says Keltner. 'The voice in your head, self-interest, self-consciousness, disappears. Here's an emotion that knocks out a really important part of our identity.' As a result, he says, we feel more connected to bigger collectives and groups.

The notion of transcending the self has traditionally been associated with religious or mystical experiences. 'Immenseness, infinitude, indescribability are some of the classical character-istics of mystical experiences that leave a person with a very powerful sense of awe,' says neuroscientist Andrew Newberg at the University of Pennsylvania, who studies how religion affects the brain. For Keltner, this is one reason why awe was so little studied until recently. 'People felt like awe is really about religion and psychologists were loath to study religion,' he says. But after interviewing thousands of people around the world about their experiences, he believes it's a mistake to see awe as inseparable from God. 'Even in really religious countries, people are mainly feeling awe in response to other great people and nature,' he says. 'People have always felt awe about non-religious things. It's available to atheists in full force.' Newberg, who is studying the awe felt by astronauts, agrees. 'You don't have to have any given belief system in order to have these experiences,' he says.

Instead, Keltner believes that awe predates religion by millions of years. Evolution-related ideas are tough to back up, but he argues that responding to powerful forces in nature and in society through group bonding would have had survival value. Chimps show signs of awe, such as goose-bumps, during thunderstorms, he notes. 'I think the central idea of awe is to quiet self-interest for a moment and to fold us into the social collective.'

It's an instinct that has been co-opted for political ends throughout history, for example in grandiose structures, from the pyramids of Egypt to St Peter's Basilica in Vatican City, or even Trump Tower. 'Awesome art and architecture have long been part of the apparatus by which people have been controlled, both socially and psychologically, and kept in their place,' says Benjamin Smith, an expert in rock art at the University of Western Australia. 'The finding that awe dimin-ishes our sense of self fits perfectly with this history.'

Despite these darker associations, there's mounting evidence that feeling awe also has personal benefits. First, focusing on the bigger picture rather than our own concerns seems a powerful way to improve health and quality of life. Keltner's team has found that feeling awe makes people happier and less stressed, even weeks later, and that it assists the immune system by cutting the production of cytokines, which promote inflammation. Meanwhile, a team from Arizona State University found that awe activates the parasympathetic nervous system, which works to calm the fight or flight response. Researchers at Stanford University discovered that experiencing awe made people feel as if they had more time – and made them more willing to give up their time to help others.

Awe also seems to help us break habitual patterns of thinking. The Arizona team discovered that after experiencing awe, people were better able to remember the details of a short story. Usually, our memories are coloured by our expectations and assumptions, but awe reduces this tendency, improving our focus on what's actually happening. Researchers have also reported increases in curiosity and creativity. In one study, after viewing images of Earth, volunteers came up with more original examples in tests, found greater interest in abstract paintings and persisted longer on difficult puzzles, compared with controls.

In the modern world, though, we're more likely to be gazing at our smartphones than at giant redwoods or a starry sky. And Keltner is concerned about the impact of our increasing disconnection from nature, one of the most potent sources of awe. 'I'm struck by how awe makes us humble and charitable,' he says. 'Is that why we have so much incivility and hatred right now in the US? I think we should be asking these questions.'

Keltner warns of a lack of opportunities for awe in poor

communities, as well as education, with its focus on test results rather than exploration. 'We are taking that away from our kids and that is a very serious problem.'

Kenneth Tupper, a philosopher of education at the University of British Columbia, agrees. 'The institution of modern schooling is very well designed to not evoke experiences of wonder and awe,' he says. This can leave teenagers feeling 'jaded and disenchanted', without a sense of connection to anything larger than themselves. To counter such alienation, he suggests, self-obsessed Western societies might consider an unconventional way to rekindle awe, taking a lesson from traditional societies. Many of these use plant and fungus-based psychedelic drugs such as ayahuasca, peyote and psilocybin mushrooms to expand the mind and forge a connection to something bigger than the self, he notes. 'These kinds of experiences are extremely highly valued.' Tupper thinks we could all benefit from similar rituals.

That's not as crazy as it might sound, according to Robin Carhart-Harris at Imperial College London. Through brain scanning, he and others have found that psychedelic drugs such as psilocybin and LSD reduce activity in the default mode network – just as awe does. In addition, boundaries between normally segregated bits of the brain temporarily break down, boosting creativity. Study participants who take psychedelics often describe being struck by vastness, and report an altered sense of self – to the point where it may disappear completely. 'My feeling is that it's the same thing,' says Carhart-Harris. 'Psychedelics are hijacking a natural system and fast-tracking people to these experiences of awe.'

There's growing interest in using psychedelics to treat anxiety and depression, but Carhart-Harris argues that if taken in a safe and controlled environment, a dose of psychedelic awe could benefit healthy people too. 'You can be more well,' he says. 'You can just feel calm and content and integrated and connected.' This idea gains support from trials of more than 100 healthy volunteers. Roland Griffiths and his colleagues at Johns Hopkins University in Baltimore found that those who took a single dose of psilocybin rather than a placebo reported feeling happier and more altruistic afterwards. They still had higher well-being and life satisfaction more than a year later.

Keltner says this is important work. 'Psilocybin should not be stigmatised,' he says. It's a potent source of awe, but there are plenty of other ways you can increase your awe quotient, he adds. First, you should raise your expectations. Put aside the myth that awe is rare, says Keltner. His surveys reveal that people feel low-level awe on average a couple of times a week. Then, think about what you find awe-inspiring. Everyone is different, but whatever does it for you, try to make it part of your everyday experience: when you're choosing which route to walk to work, which book to read or what movie to see. 'Don't think it takes big bang conversions to get five minutes of awe,' he says. 'Find your sources and go get it.'

# Acknowledgements

A huge number of people risked blowing their own minds in the creation of this book: the writers whose stories we've selected, their editors, Chris Simms and the sub-editors' desk, Graham Lawton and Sumit Paul-Choudhury, Kate Craigie, Georgina Laycock and Nick Davies at John Murray, the entire *New Scientist* family, and of course, our faithful readers.

# About the Contributors

**Anil Ananthaswamy** is a *New Scientist* consultant and author of *The Edge of Physics* and *The Man Who Wasn't There*. He teaches science journalism at the National Centre for Biological Sciences in Bangalore, India and is a guest editor at the University of California Santa Cruz's science writing program.

**Sally Adee** is a science and technology writer and editor based in London. She has reported from DARPA headquarters, the Estonian cloud and inside the young blood hype machine. She prefers a daily multi-vitamin.

**Gilead Amit** is a quondam physicist and current features editor at *New Scientist*. He knows what you're thinking and isn't in the least bit offended.

**Jacob Aron** is Analysis Editor at *New Scientist*. When he's not writing about infinity degrees Celsius he covers what's hot in the news this week.

**Colin Barras** is a *New Scientist* consultant. He holds a PhD in palaeontology and writes regularly on human evolution and the life sciences.

**Stephen Battersby** is a *New Scientist* consultant who writes popular science articles about planets, cosmology and almost everything else.

**Rebecca Boyle** is an award-winning freelance journalist in Saint Louis, Missouri.

**Catherine Brahic** is a features editor at *New Scientist*. She writes about the origins of life, the end of civilisation and everything in between.

**Michael Brooks** is a *New Scientist* consultant, and the author of several books including the bestselling *13 Things That Don't Make Sense*. He is, he freely admits, addicted to having his mind blown by science.

**Julia Brown** is a People Editor at *New Scientist*.

**Jon Cartwright** is a freelance journalist based in Bristol. He specialises in science and has a particular interest in its history, culture and effects on society.

**Matthew Chalmers** is a freelance science writer based in Bristol.

**Marcus Chown** is a science writer, journalist and broadcaster, and cosmology consultant for *New Scientist*.

**Stuart Clark** is the acclaimed author of *The Sun Kings*, is one of the UK's most widely read astronomy journalists and a *New Scientist* consultant.

**Daniel Cossins** is a features editor at *New Scientist* and generous host to innumerable microscopic sex tourists.

**Leah Crane** is a physics and space reporter at *New Scientist*. She lives in three dimensions (plus time) in Boston.

**Catherine de Lange** is deputy features editor at *New Scientist*. *Elle est particulièrement fascinée par le cerveau humain.* She's also a keen runner.

**David Deutsch and Chiara Marletto**, both at the University of Oxford, are scientists researching fundamental aspects of quantum information.

**Lesley Evans Ogden** is a freelance journalist based in Vancouver, Canada. Following a PhD in ecology, she entered the wilds of science journalism.

**Nic Fleming** is a freelance science, medical and technology journalist.

**Amanda Gefter** is a science writer specializing in fundamental physics and cosmology and author of *Trespassing on Einstein's Lawn*.

**Arran Frood** is a science journalist based in Bristol. He likes to be open minded, but not so open minded that his brains fall out.

**Garry Hamilton** is a freelance journalist whose articles have appeared in magazines around the world, including *Wildlife Conservation, New Scientist, Equinox, Audubon* and *Canadian Geographic*.

**Will Douglas Heaven** is a freelance science writer with a background in computing research.

**Joshua Howgego** has been editing features for *New Scientist* since 2015. His specialist subject is organic chemistry, which blows even his mind – and not always in a good way.

**Diana Kwon** is a freelance science journalist based in Berlin, Germany.

**Graham Lawton** is deputy editor of *New Scientist*. He has a BSc in biochemistry and MSc in science communication, both from Imperial College, and has worked at *New Scientist* for the entire 21st century.

**Simon Makin** is a freelance science journalist in London specialising in psychology, neuroscience, mental health – brains, basically.

**Jo Marchant**, PhD, is an award-winning science journalist, author and speaker. Her most recent book, *Cure: A Journey Into the Science of Mind Over Body,* was a *New York Times* bestseller.

**Michael Marshall** is an elusive freelance science journalist sighted mostly in Devon, UK, where he writes about life sciences.

**James Mitchell Crow** is an organic chemist by training, who began his science journalism career in 2007 at *Chemistry World* before joining *New Scientist* in 2009.

**Katia Moskvitch** is an award-winning science and technology journalist. A former staff writer for BBC *News*, her work has been appeared on the pages of *The Economist*, *Wired*, *New Scientist*, *Science*, *Nature*, and more.

**Tiffany O'Callaghan** is a features editor at New Scientist.

After years in advertising and ephemeral day-time television (making it, not watching it), **Sean O'Neill** wanted to hook people's attention for a nobler reason. He edits the *New Scientist* People pages.

**Stephen Ornes** is a science writer who covers math, physics, astronomy and cancer research. He works from a converted office shed in his backyard in Nashville, Tennessee.

**Joshua Sokol** is an award-winning freelance science journalist in Boston with broad interests in natural history.

**Colin Stuart** is author of six books and over 100 popular science articles on astronomy and related topics, and has an asteroid named in his honour.

**Helen Thomson** is science writer from London and author of *Unthinkable: An Extraordinary Journey through the World's Strangest Brains*.

In a former life **Richard Webb** was a researcher at the particle physics centre CERN in Geneva, Switzerland. Recognising the limits of the physics genius inside his brain, he swapped to become a journalist and is currently *New Scientist*'s chief features editor.

**Caroline Williams** is a science journalist with a fascination for the weird ways of the human mind, including her own. She recently attempted to straighten it out in her book *Override: My Quest to Go Beyond Brain Training and Take Control of my Mind*, with varying degrees of success.

**Sam Wong** is a multimedia reporter at *New Scientist*. He gets his kicks writing stories about food, drugs and animals.

*This Book Will Blow Your Mind* is edited by **Frank Swain**, Communities Editor at *New Scientist*. As well as collecting fruitloopery for the long-running Feedback column, he writes about disability, technology and cyborgs.

# Further Reading

*Cure: A Journey into the Science of Mind Over Body* by Jo
Marchant (Canongate, 2016)

*Superhuman: Life at the Extremes of Mental and Physical
Ability* by Rowan Hooper (Little, Brown, 2018)

*Factfulness: Ten Reasons We're Wrong About The World -
And Why Things Are Better Than You Think* by Hans
Rosling (Sceptre, 2018)

*Unthinkable: An Extraordinary Journey Through the World's
Strangest Brains* by Helen Thomson (John Murray, 2018)

*Reality Is Not What It Seems: The Journey to Quantum
Gravity* by Carlo Rovelli (Penguin, 2016)

*Man Who Wasn't There: Tales From The Edge Of The Self*
by Anil Ananthaswamy (Penguin, 2016)

*I Contain Multitudes: The Microbes Within Us and a
Grander View of Life* by Ed Yong (Bodley Head 2016)

*The Quantum Astrologer's Handbook: A History of the
Renaissance Mathematics that Birthed Imaginary
Numbers, Probability and the New Physics of the Universe*
by Michael Brooks (Scribe 2017)

*Unbelievable Science: Stuff That Will Blow Your Mind* by
Colin Barras (Sterling 2018)

*The Ascent of Gravity: The Quest to Understand the Force
that Explains Everything* by Marcus Chown (W&N 2018)

*Beyond Weird* by Philip Ball (Vintage 2018)

*Trespassing on Einstein's Lawn: A Father, a Daughter, the*

*Meaning of Nothing and the Beginning of Everything* by Amanda Gefter (Rough Cut 2014)
*How to Change Your Mind: The New Science of Psychedelics* by Michael Pollan (Allen Lane 2018)

# Index